WOMEN & HORMONES

"A health unto the happy!
A fig for him who frets!
It is not raining rain to me,
It's raining violets."

Robert Loveman

AN ESSENTIAL GUIDE
TO BEING FEMALE

"Women are too much inclined to follow in the footsteps of men, to try to think as men think, to try to solve the general problems of life as men solve them. The woman is not needed to think man's thoughts. Her mission is not to enhance the masculine spirit, but to express the feminine. Hers is to create a human world by the infusion of the feminine element into all of its activities."

Margaret Sanger

by the same author

All About Childbirth

WOMEN & HORMONES

by
Alice T. MacMahon, R.N., M.P.H.

An Essential Guide
To Being Female

FAMILY PUBLICATIONS
P.O. Box 940398 / Maitland, Florida 32794-0398

For further information, write:
Family Publications
P.O. Box 940398
Maitland, Florida 32794-0398 USA

Designed by Karen McGowan Steinberg
Cover Design by Ani Holdsworth
Cover Illustration by Christian Mildh

Library of Congress Catalog Card Number: 90-81214

ISBN 0-931128-03-X

Printed in the United States of America

Contents

CONTENTS

Acknowledgments

Heartfelt thanks are graciously offered to the many people who so willingly helped me with this book:

To the physicians who read the manuscript — Susan Epley, M.D.; Calvin Peters, M.D.; Bruce Crossman, M.D.; and Thomas Stanford, M.D.

To Nancy Christiansen, Louise Franklin, Darcie MacMahon, Alexis Pugh and Tom Saunders who read the manuscript and gave me thoughtful, wise and loving suggestions.

To the women who shared their stories in Chapter 9.

To the many women who have come to the Center for Women's Medicine at Florida Hospital and who have taught me so much.

To the women I work with and with whom I share the most rewarding work I can imagine.

And to Jim — my best friend and husband of 34 years for his exceptional editorial skills and his patience and support during the writing of this book.

To all the women in my life.

Introduction

*T*here is a real need for honest, simple information about hormones — a straightforward explanation of these unseen but powerful chemicals that make us female and affect us so profoundly at different times in our lives. I became especially aware of this while planning the Center for Women's Medicine at Florida Hospital and found that the number one issue with the women we interviewed was hormones — what they are, what they do and what, if anything, we can do about it. It seems that we, as women, need a feeling of control over our lives, and that this feeling of control is what is often absent during hormonal changes.

As the director of this large, very active women's center, I interact with women every day who have questions and concerns about hormones: the woman who suffers from premenstrual syndrome and wonders what causes a cheerful, agreeable person to go on a monthly roller-coaster ride of unexplained mood swings and tantrums, the woman entering menopause who worries about painful intercourse caused by vaginal dryness, and the many, many women who have questions concerning hormone replacement therapy, osteoporosis, or other issues that are predominantly female and which frequently impact our sense of self.

Women and Hormones is intended to help you to become as informed as possible about these and other hormone-dependent states of health so that you can make decisions that are best for you — whether these decisions are lifestyle related or strictly clinical. And, even though this whole subject of hormones can be highly technical, the pages that follow focus on the non-technical interpretation of the female sex hormones and how we can be fully functioning women regardless of the state of our chemistry.

Above all, I have tried to make *Women and Hormones* practical. From the explanations of the primary hormones governing menstruation and pregnancy to the descriptions of major clinical procedures, it is how you can use this information to improve your life that has been my primary motivation. Not everyone will benefit from dietary changes, new lifestyles, or all of the explanations offered, but you can at least be aware of what is possible and decide for yourself. I wish you good luck — and good health!

Chapter 1

THE CHEMICAL WOMAN

Hormones Are Us

"You can live a lifetime and, at the end of it, know more about other people than you know about yourself."

(Beryl Markham in *West With the Night*)

*W*hat are they, these things that are both praised and cursed? Webster states that the word "hormone" is from the Greek and means "to excite". It also is defined as a substance formed in one organ of the body and carried by the bloodstream to another organ or tissue where it has a specific effect. So, in the sense that we use the word hormone today, it is a biochemically specific substance that is produced in one part of the body and which triggers, influences or "excites" an effect in another.

Sounds simple, right? But this one topic creates daily phone calls and visits to my office from women saying, "I'm not stupid, but I don't understand why such-and-such is happening to me!" And this frustration is wholly understandable because, even though much is known about the chemistry of hormones, the way they influence each of us is always an individual matter and changes throughout our lives.

Actually, the body produces more than fifty different hormones. We will focus mainly on the role of the female sex hormones, however, because our sex hormones greatly influence our lives in subtle (sometimes dramatic) ways, and they are directly responsible for many of the lifestyle and health conditions that concern women. This chapter sets the stage for the rest of the book by describing these female sex hormones — what they are, where in the body they are produced, and what their functions and effects are.

But first, we need to say that the production of the female sex hormones is cyclical with each menstrual cycle lasting approximately twenty-eight days. (There is a wide range of normalcy, however, and not all women have 28-day cycles.) These cycles are repeated from 400 to 500 times in a woman's life — build up, peak and diminution. Both physical and emotional changes occur as these hormones ebb and flow, month after month, year after year, even though no two women respond in exactly the same way.

This monthly female cycle is in many ways analogous to the lunar cycle and the tidal changes caused by the moon — high and low tide, ebb and flow. Even the length of the menstrual cycle is similar to the time it takes for the moon to become full each month. Other cultures, especially Native American, draw a close comparison between a woman's "moons" and the celestial cycle.

But to understand a woman's moons from the perspective of the female sex hormones, we must first understand what happens during the menstrual cycle from a physiological point of view. This is a complex process involving several organs including the ovaries, the hypothalamus and the pituitary gland and it's important to describe this cycle before going on to describe some of the ordinary ways hormones can effect us. So stay with this "technical" explanation for just a few pages and you'll have a good basis for understanding the rest of the book. (It gets easier.)

The pituitary gland, the pea-sized gland at the base of the brain, is governed by the hypothalamus and produces two important hormones that regulate the menstrual cycle. These hormones are FSH (follicle stimulating hormone) and LH (luteinizing hormone). More about them a few paragraphs further on.

The ovaries, located just above and to each side of the uterus, release two other important hormones — estrogen and progesterone. Both of these play a key role in the menstrual cycle and, as you will see, are going to be a primary focus later in this book.

The menstrual cycle itself is divided into two phases: the follicular phase and the luteal phase. (See Figure 1.) The follicular phase begins on day one of menstruation and lasts until ovulation on about day 14. The luteal phase is the second half of the cycle beginning after ovulation and ending when the next menstrual period begins.

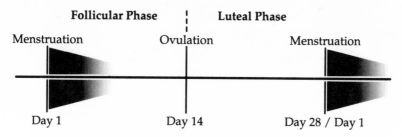

Figure 1. The 28 Day Menstrual Cycle.

If we zoom in on these two phases and look at what's happening to these glands and hormones in more detail, we will see the following sequence: during the first or follicular phase of the menstrual cycle, the hypothalamus produces a releasing hormone, which, in turn, signals the pituitary gland to release FSH. The FSH is then carried by the bloodstream to the ovaries where it stimulates the growth of an egg and a special group of cells immediately around it. These special cells, which surround the egg until ovulation, mature into a gland-like body called the follicle. The follicle itself, in turn, produces two very important additional hormones — estrogen and progesterone — which it does in varying amounts during different parts of the menstrual cycle. (See Figure 2.)

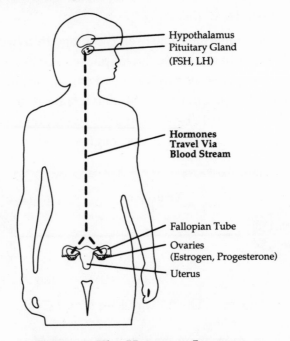

Hypothalamus
Pituitary Gland
(FSH, LH)

**Hormones
Travel Via
Blood Stream**

Fallopian Tube
Ovaries
(Estrogen, Progesterone)
Uterus

Figure 2. The Hormone Journey

Estrogen and progesterone are the primary female sex hormones and they affect us in many ways. In fact, to understand the way these hormones affect us is probably the main reason you're reading this book. But, first, it is necessary to describe their essential roles in the human reproductive process before going on to other things.

As just mentioned, estrogen is produced in the ovaries by the ripening follicle toward the early part of the menstrual cycle. This causes the lining of the uterus (the endometrium) to begin to grow. The endometrium fills with blood vessels in preparation to receive a fertilized egg, should one become available, so that it can readily supply nutrients to it during the pregnancy.

Simultaneously, the pituitary gland begins producing the hormone LH which suppresses the growth of all but one of the ovary's ripening follicles. This one follicle continues to mature until it finally releases — at about day 14 of the cycle — a fully ripened egg. This process is called ovulation, and it is at this point that the egg is ready to move into the fallopian tube and from there to the uterus. The follicle then remains in the ovary and becomes known as the corpus luteum (Latin for "yellow body").

This new phase, beginning with ovulation and lasting about 14 days, is called the luteal phase and several interesting processes start to take place simultaneously during this second part of the cycle.

As the egg is released in the ovary by the follicle, it is caught by finger-like ends of the fallopian tube. It then travels down the tube where it can be fertilized by a male sperm cell — and then into the uterus. At the same time, the follicle or corpus luteum (which surrounded the egg in the ovary), begins to produce the hormone progesterone. If the egg is fertilized, it will attach to the thickened uterine lining, the endometrium, eventually becoming a fetus, and the corpus luteum will continue to produce both estrogen and progesterone in the amounts needed for the pregnancy.

But if the egg isn't fertilized, the corpus luteum — which is still present in the ovary — stops producing estrogen and progesterone. This "tells" the endometrium that it isn't needed anymore and becomes the signal for the endometrium to leave the uterus which it does over a period of several days as the menstrual flow. Then the corpus luteum ceases to function and the uterus and ovaries return to the early follicular state ready to begin the process anew.

If conception had taken place, the corpus luteum would have remained active and would have continued to produce progesterone (literally, from the Latin, "in support of gestation") which would then be important to the progress of the pregnancy.

The timing of the above events is variable, but ovulation nearly always occurs 14 days before the next menstrual period. This ever-recurring cycle lasts an average of 28 days, whereupon the whole process begins again. It happens like clockwork with each aspect of the menstrual cycle fitting in with the others like the interlocking gears of a clock which must fit together to make the clock run.

For our purposes, then, the two main hormones to consider are estrogen and progesterone, and here is what you've been waiting for — the facts about them and the typical ways they influence us.

Estrogen

Often called the female hormone 'par excellence', estrogen functions to develop the female infant into the fully mature woman. Estrogen's main actions on the body are:

Genital Tract. Estrogen affects all areas of the genital tract, maintaining vascularity (blood supply) and stimulating growth of the endometrium (lining of the uterus). It increases the secretions of the cervix, notably at mid-cycle, making a hospitable environment for sperm.

Breasts. Estrogen stimulates development of the breasts at puberty. In adult women, estrogen seems to be the main culprit in the development of fibrocystic breast condition which causes the lumpy, bumpy breasts that plague so many women. Estrogen dilates the the ducts in the breasts, allowing fluid to collect. Then, as the estrogen level falls, these ducts constrict, trapping the fluid. This creates pressure which leads to the formation of cysts. A fibrotic process often sets in as a response to this pressure. Thus, the name describing the condition: fibrocystic. Incidentally, this is a condition, not a disease, but you will often hear it referred to as "fibrocystic breast disease".

Cardiovascular System. Estrogen relaxes the blood vessel walls causing improved circulation. This is felt to partially explain the lower incidence of heart disease in women during their estrogen-producing years.

Skeleton. Estrogen assists the bones in retaining calcium. Its diminution at menopause is felt to be a major factor in the development of osteoporosis.

Secondary Sex Characteristics. The presence of estrogen at puberty, along with the absence of the male hormone testosterone, creates the typical female body shape — narrow shoulders and waist, broad hips and the typical pattern of body hair and pitch of the voice.

Psychological Characteristics. While many factors influence our development as women, estrogen affects our moods, our zest for life and our libido. Most women report feeling more energetic, cheerful and optimistic when estrogen output is at its highest (the first half of the menstrual cycle). When estrogen production falls away and progesterone rises during the fourteen days preceding menstruation, many women report a change in mood (Chapter 3).

Progesterone

Progesterone governs the second or luteal phase of the cycle. In many ways, progesterone may be thought of as an anti-estrogen agent. Its main purpose is to prepare for and help sustain pregnancy. It is produced by the corpus luteum following ovulation as well as by the placenta during pregnancy. The actions of the hormone progesterone fall into these three main areas:

Genital Tract. Progesterone lessens cervical secretions and matures the endometrium in preparation for pregnancy.

Breasts. Progesterone stimulates the growth of the alveoli or milk-producing glands in the breasts.

Psychological Effects. In the second half of the 28-day cycle, when progesterone production is at its highest, many women report a change of mood and may feel more lethargic and depressed.

Of interest in addition to the primary hormones estrogen and progesterone, prolactin is a hormone secreted by the pituitary and is responsible for the initiation and sustaining of milk production following childbirth. It also tends to inhibit the function of the ovaries, making ovulation less likely during the time of breastfeeding. This may be the origin of the belief that you can't become pregnant while you are breastfeeding (don't count on it!). Fluctuations in prolactin levels are felt to cause water retention and subsequent symptoms related to this extra fluid, especially during the premenstrual phase.

A small amount of the male hormone androgen is produced by the adrenal glands of women. Too much androgen can cause masculinization, while not enough can create a reduced libido or sex drive. Physicians will sometimes prescribe male hormones for women to increase their libido, especially following menopause.

It is not usually the amount of female or male hormones that causes physical or psychological symptoms, but often the abrupt changes that may take place. Many women who are experiencing these problems feel they must be lacking in some particular hormone and that, if this hormone is supplemented, they will have relief of their symptoms. It is more likely the case that it is the abrupt rise and fall of hormone levels that is important in causing symptoms and not necessarily the absence or abundance of one hormone or another.

It's important, though, not to think of hormones negatively and blame them for all of the ills in our lives. Hormones seem to get a "bad rap" in our society and hormone-bashing has almost become routine.

However, as can be seen from the above explanations, our hormones do have a tremendous influence on our lives as women. From the time of puberty when the menstrual cycle begins, through the childbearing years and then into menopause, hormones are our destiny — like it or not. I'm reminded of a recent Bloom County comic strip that ended with the exasperated statement, "These women and their nutsy hormones!"

Yes, hormones are complicated and confusing. But we can benefit from understanding them, because knowing what they are and what they do helps to explain many of the changes that take place in our lives. So, come along with me on our hormone journey.

In the next three chapters, let's take a look at three specific times when hormones concern us most: when we are using hormone pills as a method of birth control, before each menstrual period when hormonal fluctuations may cause premenstrual syndrome, and at the time of menopause.

Chapter 2

THE PILL

Fooling Mother Nature

"*Death and taxes and childbirth! There's never any convenient time for any of them.*"

(Scarlett O'Hara)

"*Give me chastity and continence, but not just now.*"

(St. Augustine 354-430)

*T*he type of hormone pill that is most familiar to Americans is the birth control pill. It is so much a part of our culture that when it is referred to as simply "the pill", there is no confusion over what is meant.

The era of the pill began in 1960 when the Food and Drug Administration approved birth control pills for use following a four year clinical trial in Puerto Rico. Since then, more than 80% of American women of childbearing age have taken oral contraceptives at some time in their lives. The pill is considered to be the most effective method of contraception, except for sterilization, with a better than 99% success rate in preventing pregnancy when taken correctly and consistently.

Today's pill is much different from the early pills, primarily in the much lower amounts of both estrogen and progesterone they contain. The early pills contained more than three times as much estrogen and up to ten times as much progesterone as today's pills. The lower dosages are felt to be just as effective as the higher dose pills in preventing pregnancy and are much less likely to cause unwanted side effects. Many of the undesirable consequences of the earlier pills and their higher dosages are not experienced by women taking today's pill.

There are two types of birth control pills in use to-day: the combination pill, containing both estrogen and pro-gesterone, and the mini-pill with progesterone only. When "the pill" is referred to, it's the combination pill.

The Combination Pill. This is the most common type of birth control pill. It contains both estrogen and pro-gesterone and prevents pregnancy by suppressing ovula-tion. The synthetic hormones contained in the pill mimic the effects of natural estrogen and progesterone, tricking the body into believing that it is pregnant, thus stopping ovulation and effectively fooling Mother Nature. It also causes changes in the cervical mucous and uterine lining.

The combination pill is taken for 21 days each month. Some brands have 28 pills per pack, but the last seven are only "reminder" pills and contain no hormones.

Most women tolerate the pill very well. There are some minor side effects — most commonly weight gain, breast tenderness and nausea. These can usually be mini-mized by adjusting the type of pill taken.

The Mini Pill. The mini pill is less commonly used, but may well be the contraceptive of choice for certain women, especially those who are unable to take estrogen. The mini pill contains no estrogen, only a small amount of progesterone. It is slightly less reliable than the combina-tion pill, but the risks of complications are lower. It is taken every day, not just in a 21 day per month cycle. The proges-terone-only pill is sometimes recommended for breastfeed-ing mothers who are not candidates for birth control pills containing estrogen.

The progesterone in the mini pill alters the mucous in the cervix and makes the lining of the uterus inhospitable for implantation of the fertilized egg. In other words, if a sperm reaches an egg and fertilizes it, the mini pill will keep the fertilized egg from developing.

Side effects of the mini pill include spotting between periods, irregular periods or complete absence of menstruation.

Both of these pills, the combination pill and the mini pill, are types of hormones that are taken primarily for birth control. While different, they both have these benefits:

- The pill is effective — nearly 100% when taken correctly. The mini pill is only slightly less effective.
- The pill does not interfere with sexual spontaneity as most other methods of birth control do.
- The pill regulates the menstrual flow.
- The pill eliminates or diminishes menstrual cramps.
- The pill may help to clear up acne.

One disadvantage of the pill over less effective barrier methods, i.e., a condom, is that it provides no protection against sexually transmitted diseases.

The "Morning After" Pill. A special formulation of hormones to prevent pregnancy following unprotected intercourse has been available for several years. However, "morning after" pills should not be relied upon for routine contraception due to the increased side effects and risks that have been associated with them. They are appropriate only after a single unprotected act of sexual intercourse.

Pill Risks

There are risks associated with taking the pill and some women are not candidates to use this method of birth control. This is why it is necessary to have an assessment by a physician before you can get a prescription for the birth control pill.

Blood Clots. In some women, the pill may cause blood clots to be formed in the legs which can break away and travel to the lungs, heart or brain causing fatal pulmonary embolism, heart attack or stroke. This risk is small, but is increased in women who smoke.

High Blood Pressure. A small percentage of women who take oral contraceptives may develop high blood pressure. This effect is more common in older women or those who have a family history of high blood pressure.

Liver and Gall Bladder Disease. There is a slightly increased risk of developing liver or gall bladder disease while taking the birth control pill.

Other Risks. Other physical or mental conditions in your medical history may cause your physician to recommend against the pill as the best method for you.

While there are definite risks associated with the pill, the fact is that pregnancy itself is associated with more risks than are oral contraceptives. For most women, risks associated with the pill are considered to be minimal in comparison to risks associated with pregnancy.

Barrier methods of birth control such as condoms, diaphragms or cervical caps are often the method of choice for those who choose not to take oral contraceptives. While statistically they are not quite as effective as the pill in preventing pregnancy, these methods are wise choices when carefully and consistently used. This is especially true for the many women who distrust the pill and are uneasy using a systemic method of birth control.

Who Shouldn't Take It?

Women with a history of blood vessel diseases (such as phlebitis), women who are pregnant or breastfeeding, women with certain kinds of cancer, women with liver disease and certain other conditions (which may, in an individual case, constitute an increased risk from the birth control pill) are among those for whom the pill may not be appropriate.

Before prescribing the pill for you, your doctor will do a thorough physical examination and take your medical history to determine if there are reasons why oral contraceptives may not be suitable for you. He or she will check your blood pressure, do a pelvic and breast examination and do a Pap test, which takes a few cells from the cervix and vagina to be examined in a laboratory.

Most doctors feel that the older a woman gets, the more risks are associated with the use of the birth control pill. Women who smoke may be advised to stop taking the pill at age 35. Healthy, non-smoking women may be able to use the pill until age 40 or 45.

The effects of the pill are reversible. In other words, once you stop taking it, your body resumes its normal function. In some women, ovulation may be delayed for a few months as the body returns to normal, but long-term fertility is not changed by birth control pills. Most doctors recommend waiting for three months after taking the pill before trying to become pregnant, allowing your body time to return to normal and giving your natural cycle a chance to reestablish. During this time you should use another method of contraception.

What happens if you forget to take a pill? Usually, forgetting one pill is not a problem. Two may be. If you forget, call your doctor for directions. If you have missed more than a day or two, you will be advised to use a back-up method of contraception until your next period.

Do be aware of gaps in your pill-taking. A young couple in my childbirth class had a surprise pregnancy when, soon after the birth of their first baby, Donna had the flu. She took the pill faithfully but, after three days of vomiting, it didn't protect her against an unplanned pregnancy. It was a challenge, in more ways than one — this second pregnancy produced twins! I will never forget the Lamaze class reunion when Donna and her husband arrived with their three little boys, all under one year of age.

Some women experience "breakthrough bleeding" (vaginal bleeding) between periods. This is more common with the mini-pill than the combination pill and does not mean that the contraceptive is not working. Sometimes changing brands or dosage will correct the problem.

Safety

Concern over risks has led many women to stop taking the pill and to look into other methods of contraception, yet the pill is still felt to be the most effective method of birth control and is considered safe for most women.

Studies linking breast cancer with the birth control pill have created legitimate concern in women and their doctors. It's important to know, however, that most of these studies involved women who took the older, higher dose pill. Over 20 studies show absolutely no link between the pill and breast cancer. In fact, since the pill suppresses the natural ovarian secretion of estrogen and progesterone, some authorities feel that a woman taking it may actually be getting less estrogen than she would naturally.

The pill has not been around long enough to assess its long term effects in older women. As the pill continues to be studied over time, we will have more concrete data. Suffice to say, however, it appears today that if you have taken the pill in the past or are presently taking it, your risk of developing a major complication is small compared to the many risks associated with an undesired pregnancy.

Whether or not to take the birth control pill remains a decision you must make after carefully weighing possible risks associated with your own particlular situation.

Danger Signs

If you are taking the pill, notify your doctor immediately if you have any of the following danger signs:

- Abdominal pain (severe)
- Severe chest pain, cough, shortness of breath
- Severe headache, dizziness, weakness, numbness
- Vision loss or blurring
- Severe leg pain (calf or thigh)

Be sure to read the printed insert that comes in the package. It is required by law to be provided with all oral contraceptives to determine if you may have increased susceptibility to risks that you may not have communicated to your doctor.

While the pill appears to be the contraceptive of choice for most women today, contraceptives implanted under the skin provide long term protection without the need to remember a daily pill and may show promise for a great many women in the future.

And, while the ideal contraceptive (with no risks and only benefits) is not on the horizon, it surely will happen. Can it be any more far-fetched than space travel, computer science or quantum physics?

Chapter 3

PREMENSTRUAL SYNDROME

Life Aboard the Roller Coaster

"Due to a sudden attack of swollen feet, Cinderella suggested to the prince that he come back a bit later — say, in about two weeks!"

(PMS Attack)

*P*remenstrual Syndrome (PMS) is a serious concern to millions of women, their co-workers, their families and friends. Many women are frustrated, perplexed and at the end of their rope. So are the healthcare professionals who try to help them. And, is it any wonder, since PMS manifests differently in different women and seems to be related to more than one cause? (Could it be that PMS is nature's cure for boredom?)

The good news is that PMS, perhaps more than any other "female" problem, is extremely responsive to lifestyle changes. Several studies have shown that most women, even the "hard-core" PMS sufferers, are able to regain control over their lives by implementing the suggestions in this chapter.

Although the exact cause of PMS is unclear, the estrogen-progesterone balance that changes at the time of ovulation (at about day 14 of the menstrual cycle) is certainly felt to be a factor. Other theories that have been investigated include the body's metabolism of glucose at this time and the possibility of a defect in the way vitamin B6 is utilized. Fluctuations in prostaglandin (a hormone-like substance found in many parts of the body) may also be responsible for the monthly roller coaster ride, and a deficiency of the brain chemical, serotonin, which is important in regulation of the sleep cycle, may also be implicated. New research points to low endorphin levels as a possible contributing factor, also.

Numerous other causes have been proposed, including hypoglycemia, acid-base imbalance and chronic candidiasis (vaginal yeast infection). Until well-controlled scientific research documents the cause, or causes, of PMS, additional theories will be proposed and studied. In other words, no one knows, for sure, what causes it.

Ten to fourteen days before the onset of menstruation, most women experience physical or psychological symptoms that range from mildly annoying to those that seriously disrupt their daily lifestyle and play havoc with relationships. Symptoms such as headaches, breast tenderness, sugar cravings, weight gain, mood swings, irritability, hostility and depression are most common. More than 150 symptoms have been associated with premenstrual syndrome. In rare cases, PMS has even been linked to attacks of asthma and epilepsy, as well as suicide and homicide.

Many of the common problems it causes are psychological: depression, fatigue, tension, anxiety, feelings of hostility, confusion and bouts of uncontrolled crying. Most women report that the physical symptoms are something they can put up with, but the emotional symptoms are unendurable, both for them and for their families. It's common for women to say, "I don't know what comes over me, but every month I get totally out of control and really feel like I'm going crazy. It isn't fair to my family that I'm so hard to get along with, but I don't seem to have any control over it." This feeling of being out of control is what women describe as the worst part of their condition.

Although some form of PMS affects more than 90 percent of women, the wide range of symptoms makes it impossible to categorize PMS as a specific disease and there are no definitive tests to diagnose PMS.

Severe symptoms usually appear during a woman's 30's and often follow an interruption of the normal menstrual cycle through pregnancy, birth control pills, tubal ligation, or hysterectomy (without removal of the ovaries). And, it seems to get worse as we get older and approach menopause. This is not to say, however, that if you have trouble with PMS you will have a difficult menopause.

Many investigators have attempted to categorize PMS into groupings. Guy Abraham, M.D. identified four subgroups of PMS (or, as he states it, PMT for premenstrual tension): PMT-A, PMT-D, PMT-C and PMT-H.

Women with PMT-A experience symptoms associated with anxiety (irritability, panic, confusion, and hostility). PMT-D would cause symptoms of depression (sadness, crying, listlessness). PMT-C denotes appetite changes and cravings, often with accompanying headache, dizziness and fatigue. And PMT-H would indicate fluid retention, sore breasts, abdominal bloating and weight gain.

These categories work well in determining what the most helpful remedy may be (such as eliminating salt from your diet for PMT-H). However, since most women have symptoms from more than one category, it seldom works to focus on just one subgroup.

I recall a woman who came in to our Center for
PMS counseling recently who had a smattering of all four
subgroup symptoms. She cried, yelled and threw things,
scaring her husband and children half out of their wits. To
make matters worse, she had a constant migraine-like headache that lasted for ten days every month. The only thing
that made her feel any better at all was to eat as much chocolate as she could get her hands on. Worst of all, each month
she gained eight to ten pounds during these two weeks, ne-
cessitating two wardrobes — one for her "fat" days and the
other for when she was her normal size. Fortunately, she
was highly motivated to do something about her monthly
problem and all of her symptoms showed improvement
after she decided to follow the recommendations given later
in this chapter.

In addition to trying to determine the physical cause
for PMS, it is also helpful to look at what else is happening
in one's life. It's well known that most women are "approval junkies" — we strive to do, say, or behave in such a
way that others will always approve of us. Thus, many
women feel at some level that unless they are absolutely
perfect, they will not receive this approval.

Also, many women have been socialized to regard
their bodies in a negative way. Instead of admiring and re-
specting their bodies' reproductive functions, they feel their
monthly menstrual cycles are shameful, dirty and less than
"perfect". Take a good look at how you regard your own
body. Do you cherish it for the miraculous creation it is, or
do you find fault with the way you look, feel or experience
your life as a woman?

It's also important to realize that many women have had very negative and hurtful experiences during childhood. Sexual abuse during their early years, whether physical or psychological abuse, has tremendous influence on how these women feel about their bodies and the way they experience their womanhood. Talking with a professional counselor about unpleasant early sexual experiences can be very helpful in releasing the energy from these memories, especially as they may relate to PMS.

Documenting Your Symptoms

If you suspect that you may have PMS, the first step is to keep a menstrual calendar to see if changes coincide with the menstrual cycle.

An easy way to do this is on a calendar with large enough squares to write in (see Figure 3.). Record the first day of your menstrual period on the date it occurs, then any physical or psychological symptoms on the days they are experienced. You might code your symptoms, such as T for tension. Use a capital letter for severe symptoms and a small letter for milder ones. Write "M" and circle it for each day of your period.

Here are some suggested codes for symptoms that seem to be the most frequently reported, and a sample calendar filled out along the lines suggested. Be creative, and plug in a code for each of your major PMS symptoms on the days when you experience them.

CODE	SYMPTOM
A	Anger
AC	Abdominal Cramps
AX	Anxiety
B	Bloating
BA	Backache
BT	Breast Tenderness
C	Crying
CO	Confusion
CR	Craving
D	Depression
DC	Difficulty Concentrating
F	Fatigue
H	Headache
I	Irritability
MS	Muscle Spasms
N	Nervousness, Tension
S	Swelling of Joints (fingers, ankles)
WG	Weight Gain

These are just ideas. If you experience symptoms other than those listed, make up your own code. Remember, there are over 150 documented PMS symptoms.

Do this for at least three months. If you see a pattern where your symptoms take place between the mid-point of your cycle and the beginning of your next period, and then disappear soon after menstruation begins, it is a good indication that you are experiencing PMS. Be sure to take your symptom calendar with you if you consult your physician or healthcare provider about PMS.

Figure 3. A Typical Menstrual Calendar

SUN	MON	TUE	WED	THU	FRI	SAT
					1	2
3	4 Ⓜ AC	5 Ⓜ AC	6 Ⓜ	7 Ⓜ	8	9
10	11	12	13	14	15	16
17	18	19	20	21	22	23
24 31 AX,N	25	26 A	27 A,C	28 C,H	29 C,D	30 AX,N

What Can Be Done?

Studies show that over 80% of women with severe symptoms of PMS improve markedly by following a three-pronged self-help approach which focuses on diet, exercise and stress management. These suggestions sound so simple that you may not think they will work. But most women who try them experience very positive results from these lifestyle changes.

The only caution I would add is, if you decide to make these changes, do it consistently for a period of two or three months. And, don't cheat! They may be difficult at first, but making these changes will become much easier when you begin to see how you benefit from them. Here they are in detail.

Diet

Women report excellent, often dramatic results by avoiding certain foods during the two weeks before the menstrual period begins, particularly foods containing caffeine, sugar and salt. If the following explanations seem to apply to you, try cutting down or eliminating these items altogether. I can guarantee the "experiment" is well worth the effort for a few cycles, at least until you can identify the changes worth making a permanent part of your life.

Caffeine. Caffeine, or caffeine-like substances, is found in many beverages such as coffee, tea and soft drinks. It is also found in chocolate and in many over-the-counter medications. Even decaffeinated coffee contains a small amount of caffeine, often enough to tip the scales for someone who is particularly sensitive.

Caffeine may make you jittery and anxious and may cause mood swings. It may also interfere with carbohydrate metabolism by depleting your body of vitamin B.

Substitute with beverages not containing caffeine for those two weeks. In most cases, you will notice a big improvement.

A woman I counseled recently for PMS denied having caffeine in her diet. "I never drink coffee or tea", she said. When asked what she did drink through the day, she said that she drank five or six Cokes most days. Unbeknownst to her, she was getting a good hit of caffeine which most likely explained her extreme anxiety and irritability. She was also having trouble sleeping. Is it any wonder?

Sugar. Chocolate and other sugary foods often intensify sugar cravings, causing weight gain and increasing the body's need for B-complex vitamins and minerals. Also, if your body is having trouble metabolizing glucose, eating sweets will cause the see-saw reaction of hyperglycemia and hypoglycemia with the accompanying cravings for sweets leading to a sugar binge, followed by symptoms of weakness and the "jitters". Not to mention guilt!

Salt. Especially in women who tend to retain fluid, the PMT-H person, the avoidance of salt creates near-miraculous results. Salt increases the tendency to retain fluid. It's not uncommon for women to gain several pounds during these two weeks, just in fluid weight. The elimination of salt reduces this tendency. And, just as the cells in the breasts, abdomen and thighs retain fluid, so do the cells of the brain. This contributes to headaches, confusion and difficulty in concentrating.

Alcohol. Alcohol deserves mention because it robs the body of vitamin B and minerals and disrupts carbohydrate metabolism. Alcohol also is toxic to the liver and can impair the organ's ability to metabolize hormones. Choose non-alcoholic beverages such as mineral water with a twist of lime or lemon.

Smoking. The nicotine present in tobacco increases PMS symptoms in women much like caffeine does.

Unfair, you say. Well, maybe that's true. The very things that women who suffer from PMS may crave — sweets, salty foods, caffeine and alcohol — are the very things that may intensify the problem.

Foods That May Help Relieve PMS. Complex carbohydrates such as whole grains, legumes and fresh fruits and vegetables are good dietary choices any time, but especially during these 14 days. These foods replenish the body's supply of the B vitamins and other essential nutrients. A low-fat diet based on grains and vegetables rather than on meat and dairy products will be healthier for you.

For women who want to reduce their PMS symptoms, these dietary changes will be very positive.

Supplements. Some women benefit from supplemental vitamins, especially vitamin B6. Remember that "more" is not necessarily "better" and that too much may be dangerous. Check with your physician or nutritional consultant for the appropriate amount and kind of supplements.

The amino acid L-tryptophan, classified as a food supplement and sold over the counter, has seemed to help certain women. It is felt that L-tryptophan may raise the serotonin level, thus allowing for a more restful sleep. A glass of warm milk at bedtime will often have the same effect without the possibility of side effects.

Oil of Evening Primrose (Linoleic Acid) may help to relieve breast tenderness. This is taken orally and is available over the counter.

Exercise

Exercise, beneficial at all times in our lives, is especially helpful in relieving many of the discomforts associated with premenstrual syndrome. Those already on an exercise program will attest that they feel better physically and emotionally if they exercise vigorously. Our bodies need exercise just as they need food and water. We've all experienced a lift in spirits during a brisk walk or moving to music in an exercise class. And the "runner's high" is a well-documented occurrence. Exercise stimulates the brain to produce chemicals which enhance our sense of well-being. In fact, a deficiency of endorphins, these "feel good" brain chemicals, is one theory of the cause of PMS.

If you've not experienced this, give it a try. The next time you feel down and out, put on your most comfortable shoes and head out the door. With arms swinging, walk briskly and be aware of the change in your mood. Try singing, humming or chanting while walking — a further lift for your spirits. Even if you're too tired to think this sounds like any fun at all, try it. You'll be amazed. As one of the "Women's Own Stories" in Chapter 9 relates, a new commitment to exercise turned her PMS problem around.

Many women find that an aerobic exercise program works best if modified somewhat according to the ovulation cycle. High-level aerobics are best done during the pre-ovulatory phase, those two weeks immediately following the beginning of the menstrual period. For the last two weeks of the cycle, the time when PMS occurs, it may be best to lower the level of aerobic exercise somewhat, especially if running or bouncing aggravates tender, swollen breasts.

Add more stretching and flexibility exercises. This seems to assist the cells which tend to retain fluid during this time to eliminate that excess. Bent-knee sit ups, toe touches, waist bends and many yoga exercises help to reduce abdominal bloating.

Stress Management

Stress is the reaction to the many "stressors" in our lives — things that challenge or threaten us, sometimes unconsciously. As such, stress is inevitable. Our task is not to eliminate stress, however, which would be impossible, but to learn to manage it. A thorough discussion of stress is presented in Chapter 7 — from the power of our thoughts in dealing with it to the effect of emotions and relaxation in reducing it. But, since stress is apt to play an important part in causing or aggravating PMS, some of the more specific ways this can be dealt with by the woman experiencing PMS are mentioned here. However, be sure to read Chapter 7, because that chapter offers much more information with which to work.

A stress management class is an excellent investment. Learning coping skills will help you to deal with the stressors in your life and to control the otherwise uncontrollable reactions many women have who suffer from PMS. Because, if your hormones are already stressing you from the inside, the last thing you need is a few more stresses from the outside. These can easily become the straw that breaks the camel's back if you haven't found effective ways to minimize their influence on you.

Stressors come in three forms: physical, psychological and social. Some examples of physical stressors might be excessive weight, poor eyesight, poor nutrition, tense back muscles, alcohol or drug abuse or an injury. Psychological stressors would include anger, fears, anxieties and frustration. And social stressors include yelling children, heavy traffic, ringing telephones and even Christmas preparations.

When we are confronted with one of these stressors, we respond physically as well as psychologically. The body springs into action, exhibiting the "fight or flight" syndrome. The body mobilizes, decreasing digestion and diverting blood to the muscles and brain. The muscles tense up and sugar pours into the blood to provide fuel for energy. If this continues over a long period of time or at frequent intervals, our bodies are subjected to excessive wear and tear. Chronic back problems, ulcers, high blood pressure and migraine headaches are just a few of the common results.

If we add the physical and psychological stressors experienced during PMS, it's no wonder that women are often pushed over the threshold of their normal abilities to manage the stress in their lives.

But keep in mind that, in most cases, your reaction to the stressors in your life is really up to you. In other words, you're the one who determines how you will cope in a stressful situation. Stress is often created by the way you think about things or the way you perceive an event. Two people may react very differently to the same situation. Trying to maintain a positive attitude with optimistic expectations is a good start. And, for many people, this requires a lot of practice. At first, you may feel pretty ridiculous telling yourself that things are okay when you've overslept, the children are cranky and you're late for work, the car is out of gas and your checkbook is empty. But, how you respond to these stressors is what makes the difference in learning to manage stress.

Cultivate a positive, optimistic attitude. The woman with the "cup half full" philosophy will always fare better than the one who feels her cup is half empty.

Use the suggestions in the chapter on stress (Chapter 7) and I promise you will feel an improvement. Make use of these suggestions, especially in your relationships, where most problems relating to PMS are experienced. Whether it is a problem with your partner, your children or your work, you will get results by controlling the stressors in your life.

Support during the trying times of PMS is very important. Whether your support comes from a loving and understanding partner, family member or formal support group, it is important. Your women friends may be your greatest support and your most empathetic listeners.

And, be especially good to yourself during those days when you experience PMS. I have a friend who actually sends herself flowers when she has PMS. The flowers have become such a positive event in her life that she truly looks forward to that time of the month.

Look ahead and don't take on extra tasks or responsibilities during this time that can be avoided or postponed.

Consciously eliminate as much of the stress in your life as possible. Don't invite the boss to dinner just then, or plan your daughter's wedding or finalize your family financial plan. Reducing the stress-triggers as much as possible will help to keep your symptoms manageable and will allow you to maintain a sense of being in control.

This information on self-help measures for control of PMS symptoms will be beneficial to all women suffering from this condition. If, however, further help is needed in order to manage your symptoms and regain control of your life, suggestions that may be made by your physician could include such treatments as hormone therapy, addition of vitamin B supplements, diuretics, antidepressants or other drug therapy. Low-dose birth control pills may be of benefit to some women in controlling premenstrual symptoms.

PMS or other hormonal fluctuations may trigger psychological or spiritual "emergencies". Sometimes one may be functioning just below the threshold of symptoms and along comes a hormonal imbalance that pushes one over that threshold. Help is available. Don't hesitate to seek it. A good place to look is at your local hospital's women's center, if your community is fortunate enough to have one. And a skilled counselor can help tremendously.

It's important to know, too, that PMS is often erroneously blamed for other conditions such as dysmenorrhea (painful periods), menopause, and mental illnesses such as schizophrenia or manic-depressive disorder.

Just as we were in the dark ages about dysmenorrhea a few years ago, until the discovery that prostaglandin-inhibitors are an effective treatment, so are we still fumbling in the dark for an answer to PMS.

Although promising research is being conducted, control of your life during those premenstrual days is in your hands, and learning to manage stress is an important place to start. The lifestyle changes described in this chapter and later in this book can have profound effects on your life and your relationships and you needn't buy into the PMS stereotype if you follow them.

Nearly all of the women we have counseled for PMS at the Women's Center report significant improvement within three months of making conscientious changes as outlined above. These changes, simple as they may seem, are some of the very practical ways you can regain control of your life.

Chapter 4

MENOPAUSE

A Positive Message for a Change

"The best is yet to be."

(Robert Browning)

Women are the only female beings who live long enough to experience menopause. And, until relatively recently, human females didn't live long enough to worry about it. In the late 1800's, the average age a woman lived was only 48 years. With improved nutrition and the conquest of many diseases, today's aware woman can expect to live to be nearly 80 years old. This means that she has more than one third of her life ahead after menopause. There are now more than 40 million women in the United States who are post-menopausal.

So, menopause is not an illness but a normal and natural stage of life, one of many changes which prepare us for the next stage of living. Some women have few or no symptoms. For those who do have symptoms, however, the most annoying ones are usually caused by declining levels of estrogen. In the paragraphs that follow we'll talk about how menopause develops, when to expect it, ways you may be affected by it, and some things you can do about it — some conventional and some not so conventional. But remember, menopause is not your enemy.

Menopause is not an isolated event. It is said to occur when a year has passed without menstruation. However, during the years preceding the actual cessation of menstruation, a woman's body experiences changes caused by the gradual decline of the production of estrogen. This time is referred to medically as the "climacteric".

For a period of about seven years leading up to menopause, the reproductive system gradually changes. While there are no fixed patterns, somewhere around the age of 43 ovarian function begins to decline. The ovaries produce less and less of the female hormones estrogen and progesterone. And they stop releasing eggs on a regular monthly basis, eventually ceasing to ovulate completely. The declining hormone levels cause changes in the monthly flow. This often results in irregular menstrual cycles. Some women experience heavy, "flooding" periods. Others experience ever-diminishing scant periods. And still others simply stop menstruating.

With this irregularity, it is often hard to know if another period will occur some months in the future, so it is important to use birth control for one full year following the last menstrual period to avoid the possibility of pregnancy. It's better to be safe than sorry. We've all known or heard of women who have had a "late-in-life" baby!

Somewhere between age 45 and 55, most women will experience their last menstrual period. The average age in the U.S. is 51.

For many women, menopause occurs as a result of hysterectomy. (This "surgical menopause" is discussed in detail later in this chapter.) If the ovaries are not removed during surgery, menstruation will cease but the ovaries will continue to produce estrogen and the normal symptoms of menopause — except for not having periods — will occur later, somewhere around age 50. Since women who have had a hysterectomy no longer menstruate, it may be hard to know when menopause actually takes place. The symptoms listed below, however, are likely to be experienced by women entering menopause either with or without a uterus.

Common Symptoms

Hot Flashes (or Flushes). Hot flashes are the most common complaint of menopause. Never dangerous, but always uncomfortable, approximately 75% of women experience these disturbing changes in skin temperature and pulse rates. Up to 50% of women will experience them for five years. In addition, these hot flashes or flushes often occur during sleep as "night sweats", raising havoc with sleep patterns and creating a feeling of constant tiredness and lack of energy.

Hot flashes are usually one of the first, and most annoying, symptoms of menopause. The body temperature rises suddenly, the blood vessels of the face, neck and chest dilate and the blood rushes in and turns the skin red. Sweat may pour from the body and the heart may race. I can still see my mother fanning herself in the middle of the winter and complaining of "the vapors". These flashes, annoying and embarrassing though they may be, seldom last more than a minute or two, often only a matter of seconds. However, there is wide variation in how often they come (sometimes as often as several times a day), how hard they hit and how long they are apt to last.

When the hot flash happens during sleep, you may wake up soaked with perspiration and with your night-gown and bed linens so wet that you must get up and change them. If this happens several times each night, it's no wonder that insomnia ranks high on the complaint list.

Some simple remedies that may relieve hot flashes include the following:

- Drink ice water when you feel a hot flash coming on.
- Sleep with an ice pack by your bed or a ther-mos of ice water on the night stand.
- Splash cold water on your face.
- Take a cool shower.
- Wear cotton garments, rather than synthet-ics, and sleep in cotton sheets and nighties.
- Layer your clothing during the day so you can remove heavier jackets or shirts during a hot flash.
- Avoid situations that may trigger a hot flash: vigorous exercise on hot days, eating spicy foods or drinking alcohol.

Vaginal Dryness. Another effect of the withdrawal of estrogen during menopause is vaginal dryness, often ac-companied by itching and burning and painful intercourse. (In spite of what you may have heard, regular sexual inter-course during the menopausal years may actually prevent vaginal dryness.)

These problems of hot flashes, night sweats and vaginal dryness can be greatly helped by estrogen replacement. (See Chapter Five.)

Skin Changes. Changes occur to the skin due to the reduced estrogen level. Most women notice an increased dryness of their skin and wrinkling caused by decreased elasticity in the skin's connective tissue. Age spots, caused by hormonal changes and overexposure to the sun, may occur on any part of the body. These "liver spots" are not very responsive to removal or fading and most times recur.

Urinary Tract Problems. The lining of the bladder and urethra may become thin and dry. These changes may make you more prone to urinary tract infections. The muscles of the bladder and urethra may become weakened, as well, causing a loss of bladder control, especially when you laugh, cough or exercise. This is known as stress incontinence.

An important way to improve the overall tone of the pelvic area and help to prevent stress incontinence is to do the Kegel exercise. Designed by Dr. A.M. Kegel to help women regain bladder control, the Kegel exercises are recommended for childbearing women to promote a strong pelvic floor as well as for all women entering menopause.

To do the Kegel exercise, first locate and learn to contract the muscles surrounding the vagina, urethra, and anus. To test if you are contracting the muscles adequately, do this exercise while urinating. As you begin to urinate, contract the pelvic floor muscles as strongly as you are able. You should be able to stop the flow of urine completely.

A truly "secret" exercise, the Kegel can be practiced while sitting in a car, watching television or talking on the telephone. Tighten the muscles, hold for about three seconds, and then relax. Repeat the exercise 25 or 30 times, several times a day.

The Kegel exercise is one of the most valuable exercises a woman can learn. Practiced regularly, bladder control returns in a few months.

The Kegel is also a true "sexercise". The muscle contractions cause increased circulation to the area, assisting in vaginal lubrication. It also promotes increased tone of the vagina and greater control of the pelvic floor muscles, creating greater sexual pleasure for both partners.

Change In Body Hair Patterns. Women sometimes notice a loss or thinning of pubic hair as well as some growth of facial hair. This does not occur in all women, but some will experience this new pattern of hair growth as a result of the loss of estrogen.

Depression And Anxiety. Often blamed on menopause, depression and anxiety are more commonly caused by other life changes and should not be anticipated as a likely problem.

In fact, menopause is falsely blamed for many of life's problems. Hormonal changes are only one part of what we experience as we age. One has good days and bad days, just as at other times in life. It is important to remember that not every ache and pain is associated with menopause. And, interestingly, depression is more common among women in their twenties than at menopause.

However, the middle years of a woman's life often bring painful losses. Children leave home and parents may die. One's mate may be making his last attempts to be successful in his career, often realizing that the expectations he had when he was in his twenties are now unlikely to be met. Even worse, her mate may die (the average age of widowhood is 56), or leave her, often for a younger woman. Growing older, often alone, in a youth-obsessed society is not a pleasant prospect.

Other losses are common at midlife. We may lose our uterus surgically — 40% of women have had a hysterectomy by age 40, 50% by age 65. The loss of a breast due to cancer is not rare. And, a loss of our life's work, due to retirement or replacement, can make us feel that our life is without purpose.

The best way to cope is to stay active and healthy. The "change of life" can actually be a change for the better; a "change *in* life". Many women find their middle years to be their most fulfilling, with menopause ushering in a new stage of life filled with promise and opportunities.

Freedom from earlier cares as children grow and leave the nest allows time to rekindle the romance in a marriage. Gone now are the worries about pregnancy, allowing greater freedom in lovemaking. The ending of the monthly menstrual periods creates an ease of coming and going that you may not have experienced before; travel has never been so carefree. Now, too, you have time to devote to your career or hobby. Time to learn a new skill, to study something that has always interested you, time for friends and family without the crush of earlier responsibilities.

All of this is not to say that all women can just breeze through menopause. Many of us experience severe changes, physically and emotionally, and need more help than simply a change in attitude. But knowledge is a powerful tool. Become as well informed as possible about what is happening to your body. Then consult your physician about the most appropriate and beneficial treatment program for you. Most women will have many years to live after menopause. Learn how to make them your best years.

Surgical Menopause

Until now we have been discussing the natural progress of the reproductive slowdown known as menopause. However, when a woman has a hysterectomy (remember that 40% of American women have had their uterus removed by age 40), if her ovaries have been removed as well, she experiences an immediate and profound menopause. Her estrogen supply is removed instantly; some women have reported waking from the surgery with hot flashes. Often, the symptoms of surgical menopause are more severe, since in natural menopause the ovaries continue to produce small amounts of estrogen. If the ovaries are removed, most surgeons will recommend an immediate replacement of estrogen to avoid such dramatic and unsettling reactions.

Some reasons for having a hysterectomy would include cancer of the cervix or uterine lining, endometriosis, pelvic inflammatory disease (PID), prolapse of the uterus, severe fibroid tumors or certain precancerous conditions.

In the past, it was almost routine for surgeons to remove the ovaries when doing a hysterectomy, especially in a woman who did not desire to have more children. Today, that is less common. The major reason for removing the ovaries at the time the uterus is removed is to eliminate the possibility of ovarian cancer, a serious disease which is very difficult to detect in its early, most treatable, stages. As with many things in life, one must weigh the risks and benefits of our decisions. This is one best made by an informed woman with counsel from her physician. I loved the comment made by a woman in our hysterectomy support group when explaining her reasons for keeping her ovaries: "I just wanted to keep my own chemistry set."

Is There A Male Menopause?

A male friend once remarked during a conversation about menopause, "We've been cheated again. Not even a ritual or definite sign to mark the changes in men." This is because the specific type of hormonal activity which occurs in the female menstrual cycle and which changes so abruptly at menopause does not happen in men. A decline in the production of the male hormone, testosterone, is not inevitable and fatherhood late in life is not uncommon. (Witness Charlie Chaplin, Senator Strom Thurmond, and other elderly "new dads".)

Men do experience signs of aging, however. The pelvic muscles lose tone, the testicles shrink in size and urinary irregularities may develop. A high percentage of men over the age of 50 will develop an enlarged prostate gland that can interfere with both urination and sexual function. This is so common, in fact, that it is considered by some to be a natural part of male aging.

Most "male menopause" symptoms are, however, not physical. Men, just as women, may reach midlife without having met all of their goals or achieved all that they set out to do in life. And, men, just as women, have bad days and good days — days when they're too tired or too preoccupied to be caring and romantic. The male "mid-life crisis" is a popular subject, frequently featured by the media, addressed in counseling sessions and generally blamed for a variety of woes. But, whether it is comparable to the female menopause is doubtful.

Sexuality

The biggest problem with sexuality at menopause is in the mind. If sex was good before, it will likely be good now. Don't believe the myth that sex is only for the young. A woman can stay sexually responsive and sexually active for a lifetime. As the saying goes, there is no retirement age for sex! There are, of course, some physical changes that take place at midlife for both women and men that may influence their sex lives.

Vaginal dryness is common as the level of estrogen decreases. Also, a thinning and loss of elasticity of the vaginal walls can cause pain during intercourse — and a fair amount of discomfort for your partner, too. I recall one husband stating that he felt like he was surrounded by a tube of sandpaper! K-Y Jelly, or another unscented jelly or cream, may be all that is needed to lubricate the vulva and vagina before intercourse. Vegetable oils or massage oils are also useful for lubrication. And vitamin E oil is a good healer for fragile vaginal tissues while also relieving dryness.

A word here about estrogen cream. While this female hormone may have a very beneficial effect on the vaginal tissue, it should not be used as a lubricant before lovemaking when it could also be absorbed by your partner. Estrogen cream, often prescribed for vaginal dryness, should be used only as recommended and should be left undisturbed in the vagina for several hours after application.

As the vaginal walls become thinner, the bladder is less protected during intercourse. This can cause discomfort, not to mention the feeling of a desire to urinate during lovemaking. Not too romantic! As a preventive against this and against urinary infections, empty your bladder before and after sex to keep bacteria from entering the urethra.

In addition, as estrogen and progesterone levels fall, changes occur in the size and shape of the genitals. The vulva will decrease slightly in size, as does the clitoris. The mons (fatty tissue above the clitoris) and the labia will shrink and flatten. And the uterus and ovaries will also shrink somewhat in size.

Men, too, experience changes in their sexual response as they age. Unlike women, who are no longer capable of becoming pregnant following menopause, a man does not lose his fertility and continues to produce sperm as he grows older. However, erections are not as quick or reliable and he may require more stimulation than in his earlier years. It may take longer for him to reach orgasm and he may experience a faster loss of erection following ejaculation.

Actually, this slowing down of the male creates a better matched sexual relationship with the midlife woman. More time is usually needed by both partners in each phase. As it takes a woman longer to lubricate during the arousal stage, it may now take the man longer to become erect. His erection may be held longer than during his youth and now corresponds more closely to her slower approach to orgasm.

Ongoing and open discussion between you and your partner about your sexual relationship is very important during menopause, just as it is at other times in your life. Staying connected and in tune with each other is often difficult when so much in your life is changing.

The good news is that regular sexual activity seems to be the best method of preventing sexual difficulties during midlife for both men and women. The old adage, "use it or lose it", definitely applies.

Sexual Desire

Here, as in all of life, women are very individual. If sex has been important to you as a younger woman, it is likely to be just as important after menopause.

Some women report no change in their desire for sex, while others experience either increased or decreased desire. The Stanford menopause study determined that 71% of women noticed changes in sexual interest following menopause: 48% experienced a decline in interest, 23% noticed an increased interest and 29% reported no change at all. (I'd vote for either an increased interest or no change, wouldn't you?)

Freedom from contraception and the fear of pregnancy often yield greater spontaneity and pleasure than in earlier years. Many couples now have more time, and privacy, than they did when children were home and careers were more demanding.

In some women, however, the loss of estrogen seems to bring with it a decreased libido and sex doesn't seem appealing, either physically or emotionally. In women who have undergone a marked decrease in sexual desire, use of a male hormone is sometimes recommended.

Intercourse that culminates in orgasm is not the only way to enjoy sex, however. Mutual pleasuring by embracing and caressing, with or without intercourse, is an important ingredient in expressing love. Most women relate sexuality more to the need for intimacy and for simple body contact than they do for actual penis-in-vagina intercourse.

Studies have shown that women rate hugging and caressing as highly as they rate sexual stimulation. With age, this desire for body contact can become even more urgent— the hug, the strong hand on your shoulder or the comforting feel of your hand in his.

Things That May Lower Sex Drive

Medications. As we age, we are more and more likely to be taking medications that can interfere with sexual functioning. Some of the drugs taken for high blood pressure are particularly at fault. Inform your doctor of the problem. It may be corrected by changing the dosage or switching brands of medication. Other drugs, such as tranquilizers, can also have a negative effect on sexual desire.

Alcohol. That cocktail or glass of wine that may seem to remove barriers and enhance sexual pleasure can, in excess, have just the opposite effect. Alcohol can cause decreased libido and even impotence.

Diabetes. Long-term diabetes is an important organic cause of male impotence. Studies of female diabetics have not been as conclusive.

Fatigue and Depression. It's hard to enjoy sex if you're tired or depressed. Women and men that experience depression as a part of midlife changes may find that counseling and/or antidepressant medications will improve their desire for sex. Especially for women whose early conditioning about sexuality was negative, counseling can resolve many midlife sexual problems.

Sexuality is a part of our human birthright and is available to us to be enjoyed throughout our lives. If sexual difficulty is encountered, communicate your fears and concerns with your partner and seek help when you need it from a professional. There is every reason for you to expect this important part of your life to remain vital and fun.

Chapter 5

HORMONE REPLACEMENT

Midlife Woman's #1 Question

"Nothing in life is to be feared. It is only to be understood."
(Madame Curie)

*T*here is incredible confusion in the minds of women today concerning hormone replacement therapy. More women call and visit our women's center seeking clarification of hormonal therapy for menopause than for nearly any other concern.

Women want to know if or why they should take hormones after menopause. Much of the research in recent years has been conflicting and the journals, newspapers and women's magazines seem to report different findings and recommendations. Physicians, too, have had a problem in interpreting the scientific data, much of which is contradictory, and then communicating that to their patients.

Hormone replacement for menopausal symptoms began in 1896 in Germany with crude ovarian extracts. During the 1920's, ovarian hormones were isolated and purified. In 1935, the Journal of the American Medical Association carried the first report on the use of commercially prepared estrogen to treat symptoms of menopause.

Since those early days, estrogen therapy has enjoyed both good and bad reviews. In the 1950's and '60's, it was widely used under the banner of "feminine forever", and was touted as a veritable fountain of youth. Then, research done in the 1970's pointed to a cancer-causing effect of estrogen on the lining of the uterus. Most women and physicians became very reluctant to utilize estrogen for treating normal menopause.

Studies showed that giving estrogen alone (unopposed estrogen) to a woman who still had her uterus raised her risk significantly of developing endometrial cancer. This happens because estrogen causes the uterine lining to build up, just as it does during a normal menstrual cycle, but without a sloughing off of this built-up tissue during the menstrual period.

However, later studies showed that by adding the hormone progesterone, as happens normally during the menstrual cycle, this risk is not only decreased but is actually eliminated. In fact, current investigation shows that the combination of estrogen and progesterone protects against the development of uterine cancer and may even act to prevent it.

Today, no well-informed physician will prescribe estrogen alone for a woman unless she has had her uterus removed by hysterectomy. All women with a uterus are advised to combine the estrogen and progesterone. How this combination is given will be discussed later. Suffice it to say, it is confusing to women!

Your doctor may refer to the progesterone as progestogen, progestin, or by the common brand name Provera. When Provera is combined with the common form of estrogen, Premarin, confusion reigns. It is not unusual for women to announce that they need information on "Primavera". It sounds like something straight out of an Italian cookbook!

Why Take It?

If hormone replacement therapy can cause such concerns, why take it in the first place? A 54 year old woman recently told me, "My doctor says I should start taking hormones. But I feel just fine and I don't want to start having periods again."

True, hormone replacement is not appropriate for all women in menopause. Nor do all women choose to use it. But there are many benefits, both short and long term. By replacing at least some of the estrogen lost when the ovaries cease production, the physical symptoms of menopause are greatly reduced or eliminated. Within a few days of beginning therapy, hot flashes decrease or disappear. Vaginal dryness usually improves within the first month.

But perhaps most important of all is the role of estrogen in the prevention of osteoporosis and heart disease.

Osteoporosis is an extremely disabling disease that can limit mobility, activity and even lifespan. It results from a gradual loss of bone density, usually a consequence of decreasing levels of estrogen as we age. Estrogen helps keep calcium in the bones and has proven to slow down the rapid demineralization of bones that occurs following menopause. By combining estrogen with other preventive measures, such as calcium and exercise, most women can be spared this disease.

It is felt, also, that estrogen may protect against heart disease, the number one killer of women over 50. This is done by acting on the walls of the blood vessels and by lowering cholesterol levels in the blood. Studies show that estrogen lowers total cholesterol and LDL (the "bad" cholesterol) while raising HDL (the "good" cholesterol). Estrogen may also improve bladder control. Other less well-documented results are reported by some women as it relates to the improvement of the condition of their skin and hair. Still others sense an improvement in memory and in mood.

But, estrogen replacement will not alleviate all signs and symptoms of aging. It is not a "youth-fix" and will not magically cause wrinkles to disappear or sagging breasts to lift or tummies to flatten. Estrogen will not banish unhappiness, nervousness or depression and should not be regarded as a panacea for all ills.

Is It For Everyone?

There are some women who are definitely not candidates for hormone replacement therapy. Women who have had breast cancer should not take estrogen. It is felt that, while estrogen does not cause breast cancer, it may stimulate the growth of a cancer that may already be present. (This is also why your doctor will recommend that you have a mammogram before you begin hormone therapy — to rule out any existing breast problem.)

Also, women who have had endometrial cancer, undiagnosed vaginal bleeding, recent heart attack or stroke, recurrent blood clots, or certain types of heart disease or cancer may be advised against hormone therapy. And, certainly, if you are pregnant you should not take hormones.

How Will I Feel?

Most women notice very positive results when they take estrogen, especially in the control of hot flashes and vaginal dryness. And, most women experience no side effects from estrogen replacement. However, about 10% report irritability, abdominal cramping or breast tenderness.

These symptoms can usually be taken care of by adjusting the dosage.

One of the unfortunate, and usually unwanted effects of adding progesterone to the estrogen therapy is that most women will begin having monthly menstrual periods again, sometimes with accompanying symptoms of PMS. Thus, one of the few blessings of menopause, the cessation of menstruation, is done away with by the combination therapy. And, problems with nervousness, fluid retention and depression are sometimes attributed to the addition of progesterone.

This, more than anything else, causes women to stop taking hormones. Before making that decision, speak to your doctor. While it is true that, if you have your uterus you must add progesterone to the estrogen, these effects are usually transitory and will disappear in time. Also, the monthly menstrual periods will diminish and usually disappear after several months once your body adapts to these small amounts of hormonal additions. (No, you will not still be buying tampons when you are 80.) It may be necessary to adjust both the dosage and the method of cycling the progesterone as your body adapts to the treatment.

Hormone replacement therapy may also cause an increased risk of developing gallstones, an elevation of blood pressure or, in rare cases, the formation of blood clots. These problems are rare.

Your physician will counsel you on both the risks and benefits of hormone replacement therapy. Today, most studies show that the pros of hormone replacement therapy exceed the cons for most women. Of course, treatment should always be individualized. What is right for your friend or neighbor may not be right for you.

Follow-up care from your physician is essential, not only to rule out complications or undesirable side effects, but also to allow evaluation and possible adjustment of dosage and method of cycling. Your physician will monitor your health while you are on hormones and will likely recommend regular checkups including an annual breast and pelvic exam, Pap smear, mammogram and perhaps intermittent endometrial biopsy (an in-office procedure to determine the condition of the uterine lining).

Some physicians, still alarmed by the studies in the 1970's indicating an increased risk of endometrial cancer from estrogen therapy, are reluctant to recommend hormone replacement therapy except in the most extreme cases of women at high risk of developing osteoporosis, heart disease or those whose daily lives are unacceptably difficult during menopause.

However, most physicians today feel that subsequent studies have proven the beneficial effects of the addition of progesterone and they recommend hormone replacement therapy for many, if not most, of their menopausal patients, primarily to prevent osteoporosis. Depending upon the individual woman, hormone replacement therapy may be recommended as short-term (for menopause symptom relief), long-term (a period of several years) or life-long therapy. It remains an individual decision, one carefully made between an informed woman and her physician.

Getting Started On Hormones

Before beginning to take hormones, you should have a thorough physical examination. Your doctor will also take a complete medical history and will perform a blood test. You should also have a mammogram to rule out existing breast disease.

Other tests may also be done, including an endometrial biopsy. This test is done in the doctor's office by inserting a small instrument up through the vagina into the uterus and taking a sampling of tissue. This is a test for endometrial cancer and will give the doctor a good idea of the condition of your uterus. The test, while perhaps uncomfortable, lasts only about a minute and does not require general anesthesia or hospitalization.

A bone density test (it is non-invasive and painless) may be ordered to determine the status of your bones and your risk of developing osteoporosis. In addition, a blood test to measure the level of FSH (follicle-stimulating hormone) may be done to determine if, indeed, you are in menopause. This is more likely if you have had a hysterectomy without removal of your ovaries, in which case it is more difficult to determine when menopause occurs since you will not have had menstrual periods since the surgery.

All of these tests are important to establish a "baseline" before beginning hormone therapy. Some, or all of them will be repeated throughout the time you are taking hormones.

How Is Hormone Replacement Given?

Hormones may be given in several ways: by pill, by injection, by cream, by skin patch, and by pellet or implant.

Pill. The pill is the most common form of administration and the choice of most women. As stated earlier, for those women who have a uterus, progesterone must be given along with the estrogen to prevent the risk of developing endometrial cancer. It is felt that a minimum of 0.625 milligrams per day of estrogen (often in the form of Premarin) is required to protect the bones against demineralization. The progesterone is added for a minimum of 10 days per cycle. A typical prescription might be written as follows: take **Premarin** (estrogen) 0.625 mg per day for 25 days, and **Provera** (progesterone) 10 mg per day for the last 10-13 days, beginning on day 16 or day 13 of the estrogen cycle.

This 25-day program is followed by five days of not taking any pill at all, during which time bleeding may take place as the lining of the uterus is eliminated.

But, since hormone replacement should be individualized to the woman's own needs, your physician may recommend a different schedule. For example, you may be advised to take both estrogen and progesterone Monday through Friday, and not on the weekend, or both every day of the month in smaller doses.

The method of cycling the hormones is variable, depending upon your individual circumstances and your physician's preference. If you are having problems with breast tenderness, bloating or depression, consult your doctor so that another way of administering the hormones or a different schedule of cycling the estrogen and progesterone may be suggested.

For those women who have had a hysterectomy, many physicians today feel that it is unnecessary to add progesterone and you may be advised to take estrogen alone. Again, methods and timing of the estrogen will be individualized: estrogen every day, or just Monday through Friday, or for the first 25 days of the month.

Whatever your particular regimen may be, contact your physician with any problems you are having. Hormone therapy often requires manipulation in the early months to determine the best program for each woman. A different dosage, scheduling or method of receiving estrogen may be suggested. Don't be discouraged. Ask your physician to work with you on this.

Many women worry about what happens if they forget and miss a day or two. Unlike the birth control pill, missing an occasional replacement hormone is not a cause for concern. Just resume your daily schedule.

A common question about the hormone pill is, "How is it different from the birth control pill?" A major difference is that birth control pills usually contain synthetic estrogen, which is stronger than the type of estrogen used for hormone replacement during menopause. The hormone replacement pill is called "conjugated" (compounded) estrogen.

Injection. Hormone injections, usually at monthly intervals, may be recommended for some women, especially for those who are unable or unwilling to take oral estrogen. Since it is difficult to predict or control the level of circulating hormones and to determine dose effectiveness, this form of administration is not as often prescribed as the oral form. Women who have not had a hysterectomy would still need to take progesterone in addition to the estrogen.

Cream. Vaginal creams are often recommended for the woman whose chief complaint is vaginal dryness. Vaginal changes occur rapidly, usually within one month of initiating treatment, with progressive improvement occuring during the first one to two years. Also, it is important to know that sexual activity seems to enhance this improvement. The cream is applied through a plunger in much the same way you insert a menstrual tampon into the vagina.

It is important to use the vaginal cream as prescribed, since the same risk of developing endometrial cancer will exist if the cream is overused and the woman is not taking progesterone. Some physicians prefer the use of oral estrogen for vaginal dryness since it is difficult to judge the individual absorption efficiency for vaginal creams.

Patch. This form of transdermal estrogen acts in the same way as the skin patches used to prevent seasickness — there is a slow, gradual leaching of the medication into the system. Used as a clear patch which is stuck on the skin much like a Band-Aid, this form of estrogen replacement is tolerated well by most women, especially those who find it hard to remember to take a pill. The patch is applied to either the front or back of the torso and changed twice a week. About one half of the women using the patch will experience a skin irritation, but this is usually mild.

One of the beneficial aspects of the patch is that, unlike oral estrogen, it bypasses the liver and thus has little effect on that often overtaxed organ. Since most of the research done on estrogen replacement has been done on the oral form, however, many doctors recommend the patch only for those women who cannot take other forms or are having problems with them.

Implants. Implants are not as readily available as the previously discussed forms of estrogen replacement and are often used only under research protocol. They are used as pellets inserted under the skin. Thus, if changing the dosage becomes necessary, the treatment is not as easily manipulated as oral estrogen.

Sublingual Estrogen. This form of estrogen is placed under the tongue and slowly dissolves and is absorbed. It is another form of oral estrogen and may be the method of choice for certain women.

Research to develop the ideal method of replacing hormones during menopause continues. As more and more of our population reaches midlife and the "baby boomers" enter menopause, the subject of hormone replacement will continue to receive scientific attention.

Estrogen replacement, in whatever form, does alleviate many if not most of the symptoms of menopause. More importantly, it helps to prevent osteoporosis and protects against heart disease. And, most studies indicate that estrogen replacement does not cause breast cancer.

For those women who choose to use hormone replacement therapy, a safe and proven way of administering it is available. And, for those women at risk of developing osteoporosis, research indicates that estrogen replacement, along with progesterone if the woman has a uterus, is the proven method of prevention.

What If You Can't, Or Choose Not To Take Hormones?

Hormone replacement therapy is not for everyone. Not all women want it or need it. Many women are concerned about possible health risks that have not yet been adequately documented. If you have chosen not to use hormone replacement or have been advised against it by your physician, there are other remedies for menopausal symptoms. However, none have proven to be as effective as estrogen in preventing osteoporosis.

Some physicians recommend Bellergal-S, a combination of the drugs phenobarbital, ergotamine and belladonna alkaloids. This treatment may be effective in relieving hot flashes. Clonidine, a heart drug, is also sometimes prescribed for help with hot flashes.

Other drugs may be prescribed for insomnia, headache or depression. For vaginal dryness, lubrication in the form of K-Y Jelly or other non-scented jellies or creams, will help. And some women feel that vitamins, especially vitamins E and B, decrease their symptoms.

Relief of menopause symptoms may also be achieved by the use of natural remedies such as herbs and nutritional supplements. Some experts feel that women who consume a diet based upon grains and vegetables rather than meat avoid many of the unpleasant symptoms of menopause.

Most healthfood stores can advise you about a healthy vegetarian diet as well as herbal and other natural forms of menopause management.

Whether you choose to use hormone replacement therapy or not, it's important to remember that menopause is a normal and natural part of a woman's life. For many of today's women, it's the best part!

Chapter 6

OSTEOPOROSIS

Boning Up on Aging

"... if one just keeps on walking, everything will be all right."

(Sören Kierkegaard)

*O*steoporosis — literally, porous bones — affects millions of Americans, mostly women. Known as the "silent epidemic", it strikes an estimated one in four women by the time they are 60. Bone loss through demineralization of the bones is a natural part of the aging process of both men and women. This loss of bone occurs very rapidly in a woman at the time of menopause when her estrogen level falls and with the greatest loss occurring during the first three to five years after menstruation stops.

An estimated 20 million Americans suffer from this disabling disease which causes over one million fractures each year in people 45 and older and is the 12th leading cause of death in the U.S. These numbers will only increase as more of our population ages.

Symptoms and warning signs are usually not present until the bones have become very weak and porous and have demineralized to the point of fracturing or crumbling. Often the first symptom of osteoporosis is a bone fracture, usually in the spine, hip or wrist. The common tale of the little old lady who fell and broke her hip is usually a case of the little old lady who broke her hip and then fell. This happens when the bones become so fragile that they can no longer support weight and simply break.

Spontaneous spinal fractures, sometimes called compression fractures, cause loss of height and rounded shoulders (the "dowager's hump") as the upper spine curves outward and the lower spine curves inward. At the same time, the abdomen protrudes as downward pressure is exerted by the rib cage. Eventually, the ribs could rest on the pelvic bones, compressing the chest cavity and the abdominal cavity. These vertebral fractures are very slow to heal and are, at times, very painful. They do usually heal, even without treatment, but the changes they create in the spine often cause chronic back pain.

To understand osteoporosis, you need to first understand how bones grow. You may think of bone as solid and stony, but it is not. Bone is living tissue, constantly changing as microscopic amounts of bone cells are broken down and replaced throughout life. During our early years, as our skeleton grows, this process of bone remodeling is in a predominantly building mode, with bone being made faster than it is being broken down. The depositing cells are more active, as bones are built. During young adulthood, when skeletal growth is reached, the depositing cells and the removing cells become extra active only if a bone is fractured.

In one's late twenties, the reduction of bone begins to outpace its replacement. The removing cells become more active and bone mass gradually declines. This normally gradual diminishing of bone mass can be accelerated by negative lifestyle factors such as poor diet, inadequate exercise, smoking, alcohol and caffeine. Bone loss becomes more pronounced at menopause because less estrogen is being produced as bone replacement diminishes (Figure 4).

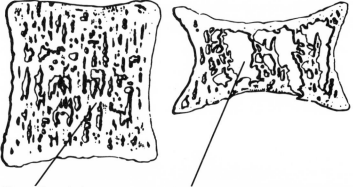

Normal bone is dense.　　Osteoporotic bone is porous and weak.

**Figure 4. Cross Sections of Normal
and Osteoporotic Bones.**

Who Is Most At Risk?

There are a number of risk factors which are believed to contribute to the likelihood of someone having osteoporosis after menopause. Here are the main ones:

- Slight build, fair skin
- Mother or grandmother with history of osteoporosis
- Smoking
- Heavy use of alcohol
- Sedentary lifestyle or chronic illness preventing regular exercise
- Early menopause or removal of the ovaries
- Certain drugs
- Fad diets, anorexia nervosa, bulimia.

Other dietary elements and some diseases can compete with and prevent the absorption of calcium. Chief among these are various compounds of the element phosphorus which are found in meat and in most soft drinks. Another culprit is aluminum which is found in some medications. Pay close attention to labels before eating or drinking them if you aren't certain of their phosophorus or aluminum content.

Some diseases also act to prevent the absorption of calcium, but the major risk factor is simply being female.

Why Are Women More Susceptible?

Women, with generally smaller bones and smaller stature, have less bone mass to lose. And, historically men have been more active than women, although that seems to be changing for today's fitness-oriented women.

But, what happens at menopause to escalate the problem? Estrogen, with its bone-protecting, calcium-conserving properties, is withdrawn as the ovaries cease production, and calcium is rapidly lost from the bones.

Men, however, do experience bone loss as well and, while it is not as common in males as in females, some men do develop osteoporosis a result of a gradual and normal physiological process. The symptoms generally appear at a later age, often not until the man is over 80 years of age. Since men have a larger bone mass than women and are not subject to abrupt hormonal interruption, they often avoid the extreme effects of bone loss. However, osteoporosis is common among alcoholic men and women of any age.

How Is It Detected?

Screening for osteoporosis can pick up early signs of the disease when treatment is most effective. However, controversy exists over the best method of testing for it and if, in fact, routine screening should even be performed.

The rationale for screening tests is to identify those women at increased risk of developing osteoporosis. This finding will also help make a determination of whether or not to begin hormone replacement or addition. Many experts suggest a bone density test at age 35 in order to have a baseline reading when the bone mass is at its peak. This can be compared to tests done at intervals thereafter as declining estrogen influences the rate of bone loss.

Since X-rays usually do not show the presence of osteoporosis in its early stages, most experts recommend one of the following methods of measuring bone density:

Single Photon Absortiometry. This method, which usually measures the density of the bones of the wrist, is less costly than other methods, but it is felt by many not to be as accurate. In the early days of osteoporosis research, the single photon method was widely used and today there are still noted experts who feel its benefits in cost and reduced radiation exposure make it the first line screening method of choice for most women.

Dual Photon Absorptiometry. This method of testing allows measurement of the density of the vertebra and hips, the sites most commonly affected. The measurement is accurate and radiation exposure is only slightly more than with the single photon.

CT. Quantitative CT (computerized tomography or "cat scan") is probably the most accurate method of assessing bone mineralization and is felt by many to represent the "gold standard" for screening. However, the cost will likely be higher than for either the single or dual photon absorptiometry and the radiation exposure is somewhat greater, about the same as that in an abdominal X-ray. Certainly, it is appropriate for the purpose of diagnosis.

Information coming from the space program and its studies of the effects of weightlessness on bone density may lead to new methods of detection. And other non-radiological methods of testing bone density are being explored.

Perhaps most important of all is the question if healthy women at low risk of developing osteoporosis should undergo routine bone density screening, given the cost factor and uncertainty over guidelines. Today, generally speaking, the jury is still out on osteoporosis screening as a routine exam for all women. For those who are making a decision about hormone replacement therapy, bone density testing is an especially helpful guideline. Many, if not most, of our osteoporosis clients are in that category. For those women at greater-than-normal risk of developing osteoporosis, screening will likely be recommended. In this instance, as in others, the decision is best made by the informed woman and her physician.

Prevention

Prevention of osteoporosis begins in childhood when bones are growing. The healthier your bones are before you reach menopause, the less likely you will be to develop osteoporosis. Tell your daughters! Awareness of osteoporosis, at an early age, is the best method of preventing its development in later life.

A well-balanced diet, high in calcium, and regular vigorous exercise during the bone growing years is like building a savings account —a bone bank. Think of it in the same way you think of a retirement or pension plan — saving now for benefit later. Unfortunately, most teenagers have just the opposite lifestyle: a poor diet, high in fat and sodium and low in vitamins and minerals is typical. And, for many teens and young adults, the extent of their exercise is getting up to change the television channel. "Bone robbers" such as smoking, alcohol and foods high in phosphorus (especially soft drinks) are much more likely to be their choices than are "bone builders" (milk and dairy products, dark green vegetables and exercise). In later life, these individuals will unfortunately enter menopause with less than adequate bone stores.

For women, prevention of osteoporosis consists of the Big Three: Calcium, Exercise and Estrogen. You can think of them as forming a three-legged stool — each leg must be of equal "length" for the stool to be balanced.

Calcium. This important mineral gives bone its structure. It not only builds strong bones and teeth, but also regulates the heartbeat and assists in blood clotting, muscle contractions and nerve transmissions. Adequate calcium intake is essential for repair and maintenance of the bones.

Your need for calcium begins before birth and continues throughout the life cycle. Calcium plays a major role in the prevention of osteoporosis and age-related fractures, so our need for this important mineral increases as we get older.

How much calcium is needed? Most authorities recommend the following amounts of dietary calcium daily:

Before menopause:	1,000 milligrams
During menopause:	1,200 milligrams
After menopause:	1,500 milligrams

If you are on estrogen therapy after menopause, you may require somewhat less than 1,500 milligrams per day. It is best to get most of this calcium from foods. However, studies show that most women consume far less than these ideal amounts. The average daily intake of calcium is 450-500 milligrams, far below what is needed by our bones.

Milk and dairy products are the best sources of calcium. Yet, many women do not like milk or are unable to tolerate it. And others avoid dairy products because they fear weight gain or an increase in cholesterol. Other calcium sources include dark green vegetables, such as broccoli. Sardines and other bony fish and yogurt are also good sources. But, as you see from the chart below, you will have to eat a lot of broccoli to get enough calcium if you are depending upon that source alone.

GOOD SOURCES OF CALCIUM

Food	Amount	Mg. Calcium
Skim milk	1 cup	300
Lowfat milk (2%)	1 cup	297
Whole milk	1 cup	290
Buttermilk	1 cup	285
Nonfat dry milk	2 Tbsp	105
Plain yogurt (whole milk)	1 cup	300
Plain yogurt (low-fat)	1 cup	400
Fruit-flavored yogurt	1 cup	345
Swiss cheese	1 ounce	270
Parmesan cheese	1 ounce	340
Mozzarella, part skim	1 ounce	210
Cheddar	1 ounce	200
Cottage cheese, 2% fat	1 cup	160
Ice cream	1/2 cup	85
Broccoli	1 cup	150
Kale	1 cup	200
Collard greens	1 cup	360
Turnip greens	1 cup	250
Bok choy	1 cup	250
Canned sardines (with bones)	4 ounces	500
Canned red salmon	4 ounces	290
Raw oysters	1 cup	225
Almonds	1/2 cup	160
Beans (cooked)	1 cup	90
Macaroni & cheese	1/2 cup	180
Cheese pizza	1/4 of 14" pie	330

In addition, many of the "sea vegetables" such as kelp contain extraordinarily high amounts of calcium and other nutrients. Get to know the foods that provide abundant calcium and build your food habits around them.

Calcium Supplements. Many women, especially those who do not consume dairy products in sufficient quantity, need to supplement their diets with additional calcium.

Supplements should be carefully chosen. Become a skilled label reader. Not all calcium supplements are of equal value. When reading the label, check for the amount of "elemental calcium". This is the amount of pure calcium contained and is usually combined with at least one other ingredient. For example, calcium carbonate may contain 40% elemental calcium. Therefore, a 500 milligram tablet of calcium carbonate would contain only 200 milligrams of elemental calcium.

Some physicians recommend certain forms of antacid tablets which are primarily calcium. Read the label to determine the amount of elemental calcium contained.

Beware, too, of the media hype surrounding calcium supplements. The public awareness of osteoporosis, positive as that may be, has also created hard-sell, scare-tactic advertising for calcium products. It has worked! Sales of calcium supplements were approximately $45 million in 1984 before the National Institute of Health issued important new information on osteoporosis. In 1985, the following year, more than $137 million in sales of calcium products were recorded.

Still the best advice is to become informed about calcium — the amount you receive each day in your diet and the amount you need. Determine the best way for you to make up the difference, whether it be through your diet, which is preferable, or through carefully chosen calcium supplements.

Vitamin D, important for the absorption of calcium in your body, is often called the "sunshine vitamin" since your body can synthesize vitamin D when exposed to the sun. Sufficient quantities of vitamin D can be obtained through good nutrition and exposure to sunlight. Some calcium supplements also contain vitamin D, but for most people, this is not necessary.

Be sure not to consume excessive quantities of vitamin D since it is fat soluble and too much can build up in your body and be harmful. The daily requirement of vitamin D is 400 I.U. Examine your diet and lifestyle (are you out-of-doors for a portion of the day most days?) to determine if you need additional amounts of vitamin D.

Excercise. The second leg of our three-legged stool to prevent osteoporosis is exercise. Exercise increases bone mass and decreases bone loss, not to mention its near-magic effect of stress reduction.

Exercise which stresses the bones, or weight-bearing exercise, is the type most effective in preventing osteoporosis. It is felt that even a daily brisk walk around the block is helpful. However, to receive optimum exercise benefit, you must exercise a minimum of 30 minutes, three to five times each week.

Walking, running, dancing, tennis or regular work-outs at a gym are all helpful. My personal belief is that, for most women, walking is the exercise "par excellence". It requires no special equipment, no monthly membership fees and is an activity that can be done lifelong. If you've not yet tried a regular walking program, I strongly recommend it. I do it myself almost daily.

Estrogen. Just exactly how estrogen acts to delay bone loss is the topic of ongoing research. One theory holds that estrogen stimulates the production of calcitonin by the thyroid gland. Another theory is that it has a direct effect on the skeleton.

In any case, recent studies have shown that estrogen is essential in the prevention of osteoporosis. Calcium alone is probably not effective in preventing bone loss, as we once thought.

The minimum dosage of estrogen for prevention of osteoporosis is reported to be 0.625 milligrams per day. However, your physician will recommend the type and amount of estrogen appropriate for your individual circumstances. As stated in Chapter 5, if you have not had a hysterectomy it is necessary to take progesterone as well as estrogen to protect the uterine lining.

Hormone therapy to prevent osteoporosis should be given as soon after menopause occurs as possible so that the rapid bone loss that typically takes place during that time does not take place. Numerous studies show that when estrogen therapy is discontinued, a rapid loss of mineral content from the bones occurs. This suggests a need for prolonged, possibly lifelong, treatment. Many physicians feel that beginning estrogen is helpful even several years after menopause. For the maximum benefit, it should always be combined with adequate intake of calcium and with exercise.

As stated earlier, hormone therapy is not appropriate for some women, especially those who have had breast cancer or certain kinds of cardio-vascular disease. For more information on hormone therapy and how estrogen is given, see Chapter 5.

Treatment

If osteoporosis is not detected in its early, most treatable stages, the bones will begin to crumble and break. Once fractures have begun to occur, treatment is designed to halt or at least to delay further bone loss and to provide support and pain relief for the injury. When fractures take place, treatment is focused on relief of pain by bracing or immobilization and by taking drugs to promote pain relief and healing of the fracture. If the hip is fractured, surgery will probably be required.

Calcitonin, a hormone manufactured by the thyroid gland, is sometimes used to treat osteoporosis. It helps to inhibit the breakdown and loss of bone mass. This drug is usually given by injection, but as nasal drops become more available, it will be easier to take.

Sodium fluoride is sometimes prescribed as treatment. This mineral stimulates bone formation. However, there are some unresolved concerns about this treatment and the side effects it may create. It is not used as a preventive, but only after fractures have occurred.

The steroid drugs Stanozolol or Winstrol (the hormones that athletes are not supposed to use and yet were made famous by one athlete who was banished from the Olympics), are possible treatment choices as are the thiazides. Parathyroid hormone is also being investigated as a treatment for osteoporosis. Undoubtedly others will be forthcoming.

For women with severe osteoporosis, vigorous exercise is not possible. Thus swimming, while not effective in the prevention of osteoporosis, is a good choice for those who already have the disease. The buoyancy of the water supports the body and it improves the cardiovascular system and one's endurance without adding stress on the skeleton.

Osteoporosis sounds scary, and is something none of us want to have. In this, as in many things in life, I'm comforted by the words of Madame Curie: "Nothing in life is to be feared. It is only to be understood."

Chapter 7

STRESS

Living in the Fast Lane

"It all depends on how we look at things, and not on how
they are themselves"

(Carl Jung)

We are all aware of the harmful effects of our stressful lifestyles. By just living at this time in history we are exposed to myriad unavoidable daily stressors: demanding jobs, traffic, loud music and blaring televisions, crime and a polluted environment. Our grandparents with their more simple lifestyles were not subject to a fast-paced life as we are today in our depersonalized society. And the effect of stress on their health was much less a problem than it is today. "Burnout" meant something totally different 50 years ago than it does today.

While stress may not actually cause disease, it certainly lowers the body's ability to resist it. Studies show that the killer T-cells which fight infections in our body are less active during periods of emotional stress and depression, lowering our immunity and allowing illness to occur.

Stress also lowers the body's ability to withstand aging. A stress-free life will not prevent the years from marching on, but the way you deal with stress will affect the way those years treat you.

Women today are especially susceptible to stress. As we have gained an increasing role in society, we are more and more affected by the formerly male-dominant stresses of the workplace. And, if we have chosen to remain in our traditional role as homemaker, we may be subject to other stressful feelings of isolation and boredom. Most commonly, we find ourselves wearing both hats: career woman and homemaker. The commonly discussed "super-woman syndrome" is often the result.

In itself, stress is neither negative nor positive. Your perception of the inevitable stressful events in your life will determine if they become your enemy. By developing coping techniques, relaxation skills and positive attitudes you can learn to manage stress and keep it from accelerating the aging process.

For most of us, stress becomes such away of life that we don't recognize it and instead walk around feeling uptight, threatened and exhausted. To see if you are under the stress gun more than you realize, ask yourself...

- Do you tire more easily than before?
- Are you working harder and accomplishing less?
- Are you increasingly cynical, sad, irritable or forgetful?
- Are you "too busy" to do even the routine things in life, and are those things falling through the cracks?
- Do you have more physical complaints than before (headaches, sore muscles, colds)?
- Does sex seem like more trouble than it's worth?

If you answered yes to any one of these questions it's time to take a good look at what's happening in your life and what you can do about it.

Stress is sometimes defined as "a product of change". The women who survive and triumph as they age are those that can adapt to change. "Openness to change" is the secret of an exciting and fulfilling life. The person who stays stuck in a comfortable rut may not experience as many disappointments as she who faces life fully and eagerly accepts risks and challenges. But the rut person will never live life to its potential. Welcome positive changes into your life. It can do much to shake loose old habits that may be creating stress.

The Power Of Our Thoughts

The power of thought is enormous. Some would say it is the most powerful force in the universe, and that out of our thoughts comes the direction of our lives. Our thoughts and beliefs determine the way we view our lives. It has been said that pain in our life is inevitable; suffering is optional. Many survivors of the Nazi Holocaust have said that they endured those horrible times by holding on to this belief.

One of the most important changes you can make in your life is to control the way you think. Negative thinking is one of our most destructive habits. We all do it, and sometimes we've been doing it for so long and are so good at it that we're not even aware that we're doing it. Step aside a moment and just observe your thoughts. Listen to what your internal "babbler" is saying.

"But, I can't control my thoughts", you say. Oh yes, you can! If negative thinking has taken over your life, call a halt to it. When the inner babbler gets going about how rotten things are, just call, "Stop". And, if it doesn't work, actually say it out loud: "STOP!" People around you may think you're a bit strange, but the babbler will learn who's boss. If it starts again, call "Stop" and don't allow the negative thoughts to escalate. Soon it will be more trouble than it's worth to think negatively and your thoughts will become more loving and peaceful. Much easier. And, much healthier.

Even with practice, though, you'll find yourself slipping back into the old negative thinking pattern. But, keep trying. As the saying goes — if at first you don't succeed, you're running about average.

We all know people who habitually think negatively. It is actually the expression of a form of fear, of insecurity. They expect a negative outcome from most situations — and, they usually get it!

If you find yourself falling into this way of thinking, try this. Instead of being against something, be *for* something. Instead of being against war, be for peace. Instead of being against violence, be for gentleness. Instead of being against poverty and hunger, be for abundance for all. Think of all the things you're against and turn them into things you're for.

Carl Jung, the great Swiss psychiatrist once said: "It all depends on how we look at things and not on how they are themselves." So true.

Negative thinking is known to suppress the immune system, not to mention ruining your life in other ways. Stop doing it — you can't afford it. Jean Houston, the popular psychologist, teacher and speaker said it so well: "At times of pathos, illness opens doors to a reality which is closed to a healthy point of view."

Attitude

Your attitude is something that you control. Just as we can learn to control our thoughts, we can also learn to change the way we perceive the stressful events in our everyday lives. Adopt an "attitude of gratitude". This may sound trite to you: be grateful for everything in your life. The wonderful things, the okay things, the so-so things and the yukkies. Be grateful for them all. The good and the bad, the dark and the light, the yin and the yang are what make up the necessary balance in our lives. If you think back to some of the most powerful and positive lessons you have learned, most have likely happened as a result of what, at the time, seemed negative and painful. It's true, we become strong at the broken places.

Be grateful, too, for all the many things we take for granted — our bodies, our food, and our planet. Be grateful for good health, good friends, water to drink, electricity, automobiles, beautiful paintings and music. There's no end to reasons for gratitude. And, it's that "attitude of gratitude" that benefits you — the grateful person. Try holding this attitude in your everyday life for a week at the minimum. I can guarantee you will be glad you tried it.

Stress Busters

Some proven ways to start managing the stress in your life instead of letting it manage you:

Substitute worry with action. The title words of the hit tune, "Don't Worry, Be Happy" could do more good than countless hours of therapy if we only believed them. Worry is negative anxiety over something that might happen and often becomes a substitute for taking action.

The worst kind of stress comes from worrying and the feeling of powerlessness that comes with it. All the "what if's" in our lives can paralyze us and keep us from moving forward. Sometimes the best motto is: ready, fire, aim. Taking action and then correcting your course if necessary is often preferable to stewing about an imagined outcome.Most chronic worriers seem to have the attitude, "Don't just do something, sit there". The opposite view as expressed by Will Rogers seems more valuable to me: "Even if you're on the right track, you'll get run over if you just sit there."

Use assertiveness skills and know how to say "No". Assertiveness is no more than defending one's own rights in a positive way. Many women see themselves in a compliant, meek role. "Don't rock the boat" is their dominant attitude. This attitude, in turn, creates resentment in oneself and a doormat way of approaching life.

It is each person's right to state and defend a position, especially as it relates to one's own life. Many women have developed the habit over the years of always needing to be agreeable, even when it means agreeing to do something that is not in their best interest. Is it necessarily in her best interest to always say "yes" to requests for help and assistance? What about the woman who works at a stressful job all week, has a family and yet feels she must volunteer, when asked, to help with the PTA carnival, collect money on her block, keep the neighbor children, or prepare dinner for her husband's client?

Nancy is a perfect example. She has a demanding job and three young children. Evenings and weekends are precious to her and her family. Yet, whenever she is asked to volunteer for any number of "good causes", she can't say no. Everyone knows they can count on Nancy. Yet, she is filled with resentment each time she agrees to something she doesn't want to do. And her stress accrues with each additional responsibility.

If any of this sounds familiar to you, I recommend an assertiveness training course. Most community colleges or women's studies groups offer such courses. You won't be sorry.

Watch self talk. How do you talk to yourself? The babbler inside your head is never silent. If you stop to listen, is it saying, "I'm not able to do this. This will never work. We'll never be able to pay all these bills. I'm too fat." Or, is the babbler saying, "What a beautiful day! I can do anything I want to. I'm capable of mastering this situation. I'm doing this because I really want to".

I especially like the last two lines of this little poem by Robert Loveman:

"A health unto the happy!

A fig for him who frets!

It is not raining rain to me,

It's raining violets."

Play often. Do you take plenty of time for yourself to have fun, or is your life totally filled with duties and responsibilities? It's well known that, in order to give to others, you must first give to yourself. If your energies are depleted, you can't energize others. As women, we've often been socialized to feel that it is selfish to take time for ourselves when so many other needs exist. We nurture others at the expense of our own nurturing.

A family therapist once did a workshop for women where he had them draw a circle and divide it up like a pie, according to how their time was spent. Work, x%, housework, y%, children, z%, and so forth. Without exception, the women all filled their circles with time spent on various activities. And not a single person had any time left for herself. His suggestion was to set aside that time first, even if it's just a few minutes each day. A good suggestion, if you keep a daily calendar or appointment book, is to write in your personal "playtime". It's just as important as all the other appointments and commitments in your life.

Taking time for yourself to have fun and do things that you enjoy is not selfish and may be the best gift you can give those close to you.

Keep a journal or notebook. Simply writing your thoughts each day keeps them from piling up and overwhelming you. Since your journal is for you alone, no one else need be aware of your feelings unless you choose to share them. Writing angry thoughts before saying them will eliminate many hostile and hurtful events, and will help you to gain perspective over what you are feeling.

Acknowledge your feelings. Feelings are neither good nor bad. They just are. It's how you act on your feelings that may be inappropriate and cause problems. When sharing your feelings with others, speak of them in "I" language: "I feel disappointed when we don't spend time together" rather than " You are never here when I need you".

Express your needs. Why is it that we expect others to automatically sense our needs? They don't. It's very unlikely that someone will say, "You look like you need a pat on the back" or "I sense that you need love". Don't expect that will happen. Instead, state your needs to those close to you. "I need some time alone", "I need reassurance", "I need help with the housework", or "I need a big hug". So often we expect our friends or family to know what we need and are then disappointed when they fail to meet those needs. Speak up, communicate your needs, then receive the results gratefully.

Form support systems. The greatest gift is to know a handful of individuals you relate to on a deeper level

than the weather, who are supportive without being judg-
mental. In our society of superficial relationships, a true
friend can be the best therapist one can have.

Focus on what's right. Instead of dwelling on what
is wrong in your life, think about what is right. Emphasize
your strengths, not your weaknesses.These words from an
old song, "Accentuate the positive, eliminate the negative,"
say it all.

Live in the moment. Most of us spend our lives focusing on either the past or the future, thinking of memories or anticipations. Our fears come from the past, which is
over, and our dreads are for the future, which may never
occur. Stay grounded in the present, in the "Holy Now".
You will find that this fraction of a moment is pretty wonderful.

Realize that you do create your own reality. If your
attitude is pessimistic, things will look very black to you. If
you are an optimist, the good things you expect from life
will happen.

Laugh or cry. In the "Health and Fitness" section of
our local newspaper recently there were two items of note
concerning good mental health and its affect on physical
health. One article recommended laughing. The other recommended crying. The laughing article began with an account of Norman Cousin's remarkable recovery from
ankylosing spondylitis by his intensive use of laughter and
positive thinking, and the fact that today many hospitals
have programs of humor therapy because of it. (Mr. Cous-
ins actually teaches his concepts to medical students now.)

Patients seem to recover from illness faster with humor in their daily lives. And there was even a recent major conference on the use of humor in pastoral counseling. In fact, there is humor to be found in even the most serious situations.

I love Woody Allen's line, "I'm not afraid to die. I just don't want to be there when it happens."

The newspaper article on crying quoted Self magazine as follows: "Crying is the oat bran of the spirit, cleanser of the soul". Tears may literally help cleanse our bodies by removing stress-related chemicals. In fact, suppressing tears may increase our susceptibility to illness.

The point is, both laughing and crying are natural responses to life situations and should be expressed when they are felt rather than repressed.

Believing is seeing. We have all been trained to know that we must first see something in order to believe it. Not so. The truth is, we must first believe it before it is real to us. If you look in the mirror and expect to see someone dumpy and unpleasant looking, that's what you will see. And, that is how you will think and behave. If, instead, you believe you will see a woman healthy and alive with a wonderful sense of purpose, that's how you will feel, act and appear to others. Try it!

The above suggestions are just a beginning and some excellent resources are available to help you, including some that are listed at the back of this book. Use these ideas in your life and I promise that you will notice an improvement. Make use of them especially in your relationships where problems are often experienced. Whether it is a situation with your partner, your children or at your work, you will get results from these "stress busters".

Relaxation

Another important way to manage stress is to learn to relax. Conscious, controlled relaxation is not at all similar to what we often think of as "relaxing". Sitting down, turning on the television and getting sleepy is not the way to learn to control your body's response to stress. Conscious relaxation is truly a life skill — one that will be helpful at all times of stress and tension in your life. And, by incorporating these relaxation techniques into your daily life, you can prevent stress from getting a grip on you.

There are two general methods of relaxation training. One is to train the body to relax first and let the mind follow. An example of this technique is progressive muscle relaxation, alternately tensing and relaxing the major muscle groups of your body. The second technique is to relax the mind first and let the body follow. This is best accomplished by mental imagery or guided fantasy. In imagery, we learn to create a positive, restful mental state, thereby reducing tension. Here are some sample scripts to illustrate both:

Progressive Relaxation.

First, get into a comfortable position, lying down or reclining, with all parts of your body well supported.

To progressively relax your muscles, begin by tensing and relaxing the major muscle groups of your body. Alternately tensing and relaxing muscles helps you to become aware of the location and feeling of muscle tension and then the contrasting feeling of the absence of tension. Tension causes pain and releasing muscle groups relieves this discomfort.

If you will be guiding yourself through the relaxation exercises, read the directions several times before beginning. If you are practicing with a partner, ask that person to read the directions to you slowly. Tense the individual muscle groups as directed, holding the tension for three to five seconds. It should begin to feel uncomfortable. Then release the tension completely, concentrating on the sensation you feel as the tension flows out. Try to tense only the muscle you are concentrating upon, keeping everything else relaxed.

> To begin, take three deep breaths and, as you exhale each time, say to yourself, "Relax, and let go".

> First, curl your toes under and feel the tension across the top of your feet. Hold it, feel the tension, then relax.

> Then, point your toes up towards your nose. Feel the tension in the back of your calves. Hold it, then relax.

> Tighten your calf muscles by pressing your heels against the floor. And relax.

Tighten your thigh muscles by pressing your knees downward. Relax.

Tense your hips and squeeze your buttocks together. Relax.

Tighten your abdominal muscles by flattening your back against the floor or bed. Relax.

Arch your back. And relax.

Now, take a deep breath and tighten your chest muscles. Relax.

Shrug your shoulders up. Relax.

Draw your shoulders forward. And relax.

Roll your head to your left shoulder. Relax.

Now, roll your head to your right shoulder. Relax.

Push your head forward to touch your chin to your chest. And, relax.

Push your head back into the pillow. Relax.

Clench your jaws tightly. Relax.

Open your mouth as wide as you can. Relax.

Squint your eyes tightly. Relax.

Wrinkle your forehead by raising your eyebrows. Relax.

Tense the muscles of your scalp. And relax.

After progressing this way through all of the muscle groups of your body, spend a few more minutes relaxing all over. Continue to feel yourself becoming more relaxed each time you exhale, with a feeling of sinking deeper into the floor or bed. Now, gradually return to full awareness of your surroundings.

Practice this progressive relaxation exercise every day. If you have difficulty staying awake, you may need to practice in a sitting position with your eyes open and concentrating on a "focal point" in the room.

A suggestion that many people find helpful is to tape the relaxation directions in your own voice and play the tape back as you practice. If you make a tape, choose a time when you are calm and quiet so your voice is without stress. Then record the directions in a soft voice. As you listen later, it will be a way to bring your frazzled self into harmony with your quiet, centered self.

Soon you will be able to use this technique to relax whenever you feel the stress of everyday life. Practice mentally scanning your body for tension throughout the day and relaxing it away as you exhale.

Imagery.

Mental imagery is a little like a daydream. It is a deliberate use of something we do all the time — use our imagination.

By developing your ability to "image" a desired situation, you will enhance your relaxation skills. Learning to create a positive, restful mental state allows you to reduce the degree of tension present during stressful events. Mental imagery helps to reduce mental anxiety and tension just as relaxation reduces physical tension.

In the beginning, you may want to practice visualizing in your mind's eye a familiar place. Close your eyes and try to visualize your bedroom. See the colors of the walls and the bed covering. Are there pictures on the walls? What do you see on top of your dresser? What other furniture is in the room? Try to actually feel being there.

Try this same technique to mentally see the interior of your car. Visualize the upholstery, the steering wheel, the dashboard. Where is the radio located? And the glove box? Concentrate deeply until a picture or sensation of the space forms in your mind and you actually begin to feel that you are right there.

Another familiar scene to practice learning to visualize might be your bathroom, seeing the wallcovering, the sink and mirror, the towels, the tub or shower tile, your toothbrush, and so on. Do the same with your kitchen.

By practicing visualizing familiar places, you will develop your creative awareness and ability to mentally enter a scene or situation of your choice.

As your mental pictures become clearer, you may also want to add your other senses. Not everyone forms a mental image pictorially. Some recreate a scene in their mind by the sounds they hear, the feel of the air on their skin or the smells they associate with the scene or event.

In the following guided imagery example, notice that all of the senses are involved. Have someone read it to you slowly as you close your eyes and relax with these words....

Take a comfortable position and close your eyes. Take three breaths as you let go of the day's tension. Relax your body from head to toe.... Imagine yourself in a green field on a warm, spring day. As you rest on the lush, green grass you notice that the field is full of clover, with occasional yellow dandelions. The leaves on the trees rustle in the gentle wind. The soft breeze blowing over your skin soothes and relaxes you. In the air are the pleasant scents of grass and flowers. Nearby are lilac bushes — the breeze carries their fragrance to your nostrils. Billowy white clouds spot the bright blue, sunny sky. A few birds circle lazily in the sky, while others are singing in the trees. You walk to an apple tree nearby and pick a bright red apple, warm from the sun. Your teeth pierce the skin of the apple and you taste the sweet, juicy fruit. You are peaceful and happy.

Were you successful in mentally traveling to this pleasant scene? Did you see the rich, green color? Did you feel the air on your skin? Did you hear the birds and smell the flowers and grass? Did you, perhaps, even begin to salivate as you mentally tasted the apple?

Try several different kinds of "mind trips". Your partner can help you to mentally travel to your favorite place — to the beach, to the mountains, to a pleasant meadow, to the hammock in your back yard — by verbally guiding you and suggesting the images you will experience. If you are practicing alone, create your image in as much detail as possible. Create your own "special place or scene to visit; you will find that certain scenes are more conducive to relaxation than others. Choose the scene that works best for you.

You may find that background music helps tune out distracting thoughts and sounds, music that is nonintrusive and simply pleasant and soothing. Instrumental music is usually less distracting than vocals, provided that the sound is even and pleasant. So-called "new age" music lends itself very well to relaxation, as do recordings of environmental sounds. There are also a number of commercially recorded tapes produced for an optimum balance of sound, music and suggestion for relaxation. Usually these can be found in music and book stores. The preferred way to use them is with lightweight earphones.

Don't be discouraged if unwanted thoughts intrude as you begin mental imagery. Be patient. Complete concentration requires a great deal of practice. You will find that the more adept you become at focusing your attention on your mental image, the less you will be aware of these random thoughts. Also, both diet and exercise are proven stress reducers. They are covered in Chapter 8.

The above suggestions are just a beginning. There are many excellent books which are totally devoted to improving one's attitude and creating a positive outcome from seemingly negative life situations. Signing up for a stress management course is an excellent investment. And, tuck away a few little "gems" to call on when things get tough. Two of my favorites are:

Forgive the past and let it go — for it is gone.

(A Course in Miracles)

Follow your bliss.

(Joseph Campbell)

Chapter 8

STAYING YOUNG

Growing with the Flow

"How old would you be if you didn't know how old you were?"

(Satchel Paige)

taying young has a lot more to do with attitude and lifestyle than with any predetermined factors like heredity. How you look, act and feel is determined by many factors, over most of which you can exercise a great deal of control. How old you are is determined in large part by how you balance your physical lifestyle and your psychological lifestyle.

Who you are means the combination of many elements that makes you unique. Your attitude, your mental abilities, your humor, the way you express yourself, the love you show others. These qualities all make up your character. The way you look is only one part of your "gestalt", the image people have when they relate to you.

This chapter will cover how we can stay young, no matter what our age. We will look at the common characteristics of youthful people, how to "mind" our bodies, the tremendous benefits of proper nutrition and exercise, and how to care for our skin — the package we present to the world. We will also look at the various types of medical and surgical skin treatments that can be used to rewrap that package.

The more we learn about how to keep our bodies and minds young, the more control we have over our own successful aging. Exercising, eating a balanced diet with enough calcium and fiber, drinking plenty of water, taking care of our teeth, not smoking and protection from the sun can actually ward off old age as well as minimize the risk of developing heart disease, osteoporosis, certain forms of cancer and even wrinkles.

Today's midlife woman has many healthy and beautiful role models to follow. I think immediately of Jane Fonda, Shirley MacLaine and Gloria Steinem. All are over fifty years old and all have made a real impact on our society and the way we think about women as they age. No one thinks of these women as old or declining!

Jacqueline Kennedy Onassis is in her 60's. Do we think of her as "an old lady"? Never!

And who could think of Tina Turner with her boundless energy as old?

My favorite role model is my own mother-in-law who, at age 87, is as vital and alive as she has ever been. Not only is she informed on the affairs of the world and its inhabitants, but she is totally independent, driving her car wherever she wants to go. She has even taken on a weekly volunteer job that requires her to use the computer! Her sense of excitement and romance is as strong and alive as when I first met her nearly forty years ago. We joke about how I would love to have her genes — which, of course, I can't. Her secret, though, is her attitude. And that I *can* aspire to!

A motto for successful aging might be: "Good genes, good habits, good attitude and good luck!

Characteristics Of Youthful People

People who seem to stay young, no matter what their age, have several characteristics in common: they don't accept society's cultural expectations and they reject labels. They almost always think of themselves as younger than their "real" age. They have a sense of purpose, whether it be a job, their family or their community. They have a sense of being needed and of making a contribution. They are interested, alert and always focused on a goal. They live "in the moment" instead of looking backward or forward.

These people stay mentally active throughout their lives, oftentimes by such "ordinary" practices as doing the daily crossword puzzle in the newspaper. They are constantly learning by enrolling in classes, attending community programs or traveling. Their friends who age less successfully tend to be passive learners, often absorbing what they learn from television. These passive learners feel that the end of school is the end of their education. They are more idle in their free time than their "ageless" friends. And, they don't maintain the same clarity of thought.

Youthful people, whatever their age, have nutritious eating habits. They eat foods that are high in the nutrients that keep the body strong and lean. They eat regularly, starting each day with a healthy breakfast. They eat red meat sparingly, stay away from drinks and snacks high in salt, fat and sugar. And they avoid fad diets.

Exercise is a natural part of daily activities for these people. They also have a planned exercise program to which they are committed and rarely skip, even if it means re-scheduling something or running in the rain.

People who age successfully take care of their bod-ies. They avoid toxins and environmental contaminants as much as possible. And, they do not smoke. Smoking may be the single most important change people can make who want to remain healthy throughout their lives. Smoking dries the skin, creates wrinkles and is a known cause of many diseases.

Another characteristic of youthful people, of any age, is that they have loving, supportive relationships with oth-ers: family, friends, colleagues, even pets. None of us lives in isolation, even though we sometimes feel that we do. The more "strokes" we give to and receive from others, the more energized and vital we become.

Youthful people are optimistic, viewing adversities as challenges. They are innovative problem solvers. They do not blame others or themselves for what happens in life and regard problems as simply mistakes and learning op-portunities. They have a sense of humor and know how to play. They are open, enthusiastic and fun to be with. They have not lost touch with the child within themselves, and they are creative rather than competitive. They focus on exploring and enjoying the fullness of their own gifts rather than comparing their gifts and attributes to those of others.

And, lastly, youthful people all have a healthy li-bido and a positive attitude toward their sexuality. People who feel old report a decline in their interest in sex, whereas youthful people stay sexually active and engage in regular lovemaking no matter how old they are. As a professional friend of mine is fond of saying, "It's the last thing to go!"

It is definitely not true that life is all downhill after fifty. As Victor Hugo wrote, "Each half of life has its youth and its old age: forty is the old age of youth, fifty is the youth of old age". Since Victor Hugo's day, we have increased our life expectancy well beyond this.

On a somewhat less positive note, one of the myths of midlife is that of the woman sending her husband off to work while she stays in an empty house writing letters to grown children and reminiscing with old photo albums.

Today well over half of midlife women are employed outside the home due either to choice or circumstance. Being single at midlife is commonplace. One out of three marriages ends in divorce, and many midlife women become widows. Being totally responsible for herself financially is common for today's woman of any age.

Many women must re-enter, or enter for the first time the job market, which is not always easy. Fortunately, displaced homemaker programs, job search clubs and career counseling are available in many communities. Your local community college or chamber of commerce can help.

Minding The Body

In addition to adopting a more healthy lifestyle, which this chapter is about, there are important ways to monitor your health and to practice prevention. Medical counsel is always important, but ultimately the responsibility is yours.

Your Breasts. Early detection of breast cancer is of major importance. While the statistics on this disease are alarming — one in ten women in the United States will develop breast cancer at some time in her life — the good news is that, by finding breast cancer early when it is very small, it is not only treatable, but there are many more options available for that treatment. The three best defenses against breast cancer are personal physician care, mammography and breast self-examination.

1. Personal Physician Care. Through regular physican examinations, your physician is able to monitor your health and can alert you to potential problems before they become serious.

2. Mammography. A mammogram (a low-dose X-ray of the breast) can often detect breast cancer long before it can be felt. Many women avoid having this important test due to fears and misunderstandings about the procedure. These fears include a fear of radiation, but today's modern equipment, unlike that of several years ago, emits a very low amount of radiation and is considered to be extremely safe. In fact, it is felt that you would need to have 300 mammograms to raise your risk of developing cancer from radiation exposure by 1%. Also, fear of what will be found is common, but this "ostrich" attitude is not only dangerous, but unreasonable. Actually, most problems found by mammography are not cancerous, and those that are can be much more easily and successfully treated if they are found at an early stage. And, finally, fear of pain puts off some women; there may be slight discomfort from the necessary compression of your breasts during a mammogram, but it should never be painful. It is best to schedule your mammogram after your menstrual period, rather than before, if you tend to have breast tenderness premenstrually.

3. Breast Self Examination (BSE). Most breast lumps are discovered by women themselves. A well-trained woman, who knows her own body and the way normal breast tissue feels, can detect breast lumps that are abnormal while they are still very small. Every woman over the age of 20 should examine her breasts every month in the following three ways:

1. In the Shower. Raise one arm. With fingers flat, touch every part of each breast, gently feeling for a lump or thickening. Use your right hand to examine your left breast, your left hand to examine your right breast.

2. Before a Mirror. With your hands on your hips, look carefully for changes in the size, shape, and contour of each breast. Look for puckering, dimpling, or changes in skin texture. Do the same with your arms raised above your head. Next gently squeeze each nipple and look for discharge.

3. Lying Down. Place a pillow under your right shoulder and your right hand behind your head. Examine your right breast with your left hand. With fingers flat, press gently in small circles, starting at the outermost top edge of your breast and spiral in toward the nipple. Examine every

part of the breast. Then repeat with the left breast, changing the position of the pillow and using the opposite hand.

The American Cancer Society offers the following guidelines for the early detection of breast cancer:

- Women 20 years of age and older should perform breast self examination monthly.
- Women between 20 and 40 should have a physical examination of their breasts every three years.
- Women between 35 and 40 should have a baseline mammogram.
- Women 40 to 49 years of age should have a physical examination of their breasts annually, as well as a mammogram every one to two years.
- Women over 50 should have a mammogram every year when feasible.

Your Bones. As explained in the chapter on osteoporosis (Chapter 6), prevention is the key. Be certain that you are receiving adequate calcium in your diet, do regular weight-bearing exercise to strengthen your bones, and speak to your physician about the advisability of adding estrogen at the time of menopause.

Osteoporosis screening is available in most communities. Measuring the density of your bones can indicate how susceptible you may be in later years to developing this disease. The testing is completely safe and comfortable.

Your Reproductive Organs. Even though you may be past your childbearing years and do not need obstetric care or birth control, you do need regular medical checkups to determine the status of your uterus and ovaries. A physical examination, including a Pap smear and pelvic exam, can test for the presence of disease. Many women feel that this is no longer necessary once they reach a certain age, or after they have had a hysterectomy, but this is not true. We need continuing health assessments as much, if not more, during our middle years and beyond than we did during our childbearing years.

Your Heart and Blood Vessels. Our lifestyle and the wear and tear of time can have a serious effect on our cardiovascular system. It is important to know what your blood pressure is and, if it is elevated, what to do about it.

The same is true of your cholesterol level — both your total cholesterol and the amount of low-density lipoprotein (LDL or "bad" cholesterol) and high-density lipoprotein (HDL or "good" cholesterol) should be tested.

If these tests are not within normal limits, learn what you can do to correct it. As with many things in life, diet, exercise and lifestyle have an effect on your cardiovascular system and are the keys to good health.

Your Digestive system. Aside from making certain that your diet includes that special friend to your colon, fiber, it is recommended that you take the following precautions against colon cancer:

1. A rectal exam by your physician every year after 40.

2. A stool blood test (hemoccult) every year after 50.

3. Sigmoidoscopy (a visual inspection of the interior of the colon using a flexible "scope") every three to five years after 50.

Nutrition And Healthy Eating

The old saying, "You are what you eat" is largely true. Proper nutrition is essential in managing the aging process. The right foods will help keep our bones and teeth strong, will keep your heart and lungs operating efficiently, will maintain the tone of your skin, will keep your mind sharp and alert, and will help you to resist disease. The wrong foods add toxins to the system and deplete essential nutrients. High fat, high sugar foods contribute to aging.

Your primary source of nutrition must come from the foods you eat. Don't count on nutritional supplements to provide what your body needs. Here are some guidelines for maintaining a healthy dietary program.

Eat a variety of foods each day. Instead of focusing on "I must eat this" and "I can't eat that", include a wide variety of foods in your diet. Vary the shapes, colors and textures of the food you eat — the more interesting it is, the more you will enjoy it and the better for you it will be. Include fresh fruits and vegetables, whole grain breads and cereals, milk and milk products and protein foods such as meat, fish, poultry and dried beans.

Eat foods with adequate fiber. Fiber is essential for keeping your digestive tract in a healthy condition. Practically speaking, this means choosing natural, whole foods and avoiding processed ones.

Maintain a desirable weight. Avoid reactive, compulsive dieting and instead manage your weight by increasing your physical activity and decreasing the amount of calories you consume. There is no magic here.

Avoid fat. Choose low-fat protein sources, trimming excess fat from meat before cooking. Use low-fat or skim milk and low-fat cheeses. Broil or bake foods instead of frying them. And limit the amount of oil and dressings added to salads and other foods.

Avoid sugar. Select desserts that are lightly sweetened or not at all, such as fruit. Choose fresh or water-packed fruit. Avoid concentrated sweets like candy and soft drinks.

Avoid salt. Don't add salt in cooking and use it sparingly, if at all, at the table. Limit salty foods, avoid cured meats, and read food labels for sodium content (table salt is listed as sodium chloride). If you have been accustomed to using salt as flavoring, it will take about three weeks of eliminating salt before food will taste more "normal" again.

Avoid alcohol. If you drink alcohol, do so in moderation. More than one or two drinks per day will deplete your body of essential nutrients and damage your internal organs.

Drink plenty of water. Eight glasses a day helps eliminate toxins and creates a healthy, vibrant skin.

Emphasize grains and vegetables in your diet. It is common knowledge that many animals which are raised for human consumption are given food containing antibiotics and hormones. Many people feel that women who are vegetarians or on a grain-based diet do not experience the same level of difficulties with premenstrual syndrome or menopause as those who consume a meat-based diet.

Eat to live and to stay fit for life. We all know those who live to eat and who view life as a giant smorgasbord. Instead of using food as a reward, use food as fuel. And don't overfill your tank!

Exercise

The shape and firmness of your body when you are fifty will never be exactly as they were when you were twenty. But, how strong and lean your body looks and feels as you age is largely up to you. It is easy to blame that "matronly" look on years or on that relentless enemy, gravity. But the usual culprits are a less than ideal diet and a sedentary lifestyle.

Contrary to popular belief, our need for exercise does not decrease as we get older. Every system in our body needs exercise to remain in optimum working order. People who exercise regularly have fewer symptoms of aging than those who do not. Exercise benefits both the body and the mind, and, it's never too late to start! Hulda Crooks started mountain climbing when she was 64. She is now 93 and still climbing to the top of Mt. Whitney every year. In 1987, at age 91, she led a group of 300 people to the 12,388 foot peak of Japan's Mt. Fujiyama. Mavis Lindgren at age 82 is still running marathons. She began running in her 60's and has completed 52 marathons since she was 70 years old.

Many women who were sedentary in their earlier years and who take up exercise as a way of life actually look and feel healthier at midlife than in their twenties.

The type of exercise is up to you. Choose something you enjoy, that exercises your whole body and provides an aerobic workout to strengthen your cardiovascular system. Whether this is walking, running, tennis, dancing or bicycling, enjoy it and you will be more likely to continue.

The exercise question most often asked of me is, "What kind of exercise is best for me?" For most, walking is an excellent choice. It is always available and requires no special equipment or monthly membership fees. Best of all, it's an exercise you can enjoy for the rest of your life. There is less risk of injury from walking than from exercise that is more vigorous and which puts stress and strain on the joints. And the benefits are comparable to the more rugged forms of exercise. Brisk walking is proven to lower stress and promote healthy sleep. Add that to the improvement in the way you look and feel, who could ask for anything more?

To be most effective, aerobic exercise must be steady and sustained for 15-30 minutes and should be done a minimum of three to five times each week. This should be preceded by a warm-up of stretching and flexibility exercises to prevent injury. To round out your exercise program, the following exercises will improve muscle strength and tone:

Waist. Stand with feet parallel with one hand on your hip. Curve the other arm overhead. Bend sideways at the waist toward the hand on your hip and press gently, downward and back, for 8 counts. Repeat on the other side. Alternate left and right for 4 sets, 8 counts each.

Arms. Stand with feet parallel and arms extended to either side at shoulder level, palms facing the floor. Keeping your arms straight, make small circles forward for 8 counts, then backward for 8 counts. Repeat, making larger circles forward and backward, 8 counts each. Repeat the entire sequence 4 times.

Abdomen. Lie flat on the floor, knees bent and hands behind your head. Press your lower back against the floor. Curl up, one vertebra at a time, until your shoulders are off the floor. Return to original position. Start with 5 repetitions, work up to doing 15-20. Make your tummy muscles do the work — don't lunge up with your arms forward.

Hips and Thighs. Lie on your left side, knees bent, your weight resting on your elbow and forearm. Straighten your left leg and lift it a few inches off the floor, then lower. Do this slowly, to a count of 8. Repeat 4 times, resting in between. Then repeat on the other side.

You can also incorporate exercise into your daily life. Use the stairs instead of the elevator, park far enough from the shop door to get in some extra steps (and your car is not as likely to get dinged!), and use your feet instead of your telephone whenever possible. You will be surprised at how much physical activity you can build into your day.

If you have any physical problems, ask your physician to recommend an exercise program that will be most suitable for you.

Skin

The skin, our largest organ and the wrapping we present to the world, is meant to protect us from the environment like a coat of paint on a car. It also seals in moisture and controls the body's temperature by perspiration and evaporation. Drying, wrinkling, and sagging are inevitable, to some extent, but can be minimized and even partially reversed by following a few simple observances.

As we age, however, our skin does becomes thinner and more susceptible to damage. It is less elastic than thicker skin and tends to sag and wrinkle more.

Wrinkles occur when the deep layer of the skin, the dermis, loses moisture and elasticity. As the dermis shrinks, the skin's top layer, the epidermis, becomes too loose. The epidermis contracts, leaving behind tiny creases and folds, otherwise known as wrinkles. The extent and timing of wrinkling are dependent upon several factors: your genes; your facial habits, such as habitual frowning or pursing of the lips; and, most of all, sun exposure. The skin on your body tends to show signs of aging later than facial skin since it is more protected from the sun.

As we age, our skin takes longer to "rejuvenate". Normally, skin cells rise to the surface of the body and are sloughed off by friction. New skin is formed in about 30 days in a young person. It takes longer and longer to complete this skin renewal cycle as we age, so that an older person's skin is actually "older" than the skin of a younger person and may look duller and less fresh.

A man's skin ages somewhat slower than a woman's. Not fair! His skin is normally more oily and less susceptible to drying. And, he has one big friend that we don't — the daily shave that removes the old skin and encourages rapid growth of new skin.

Our skin has many enemies: sunlight, alcohol, tobacco, caffeine, diets, facial expressions and inappropriate use of cosmetics.

The harmful elements in tobacco create a typical "smoker's face", characterized by lines radiating from the mouth and eyes and a sallow complexion. Alcohol causes the blood vessels to lose their elasticity and to break. We've all known heavy drinkers who have red and blotchy skin, especially around the nose.

However, over exposure to the sun is the biggest cause of skin damage, producing leathery, wrinkled skin and often leading to the development of skin cancer.

One of the best habits we can develop to improve our skin is to drink plenty of water. Women achieve noticeable improvement in the condition of their skin in just a matter of weeks once they embark on the eight-glasses-a-day program.

Skin Care. To cleanse your skin, use a mild, non-abrasive soap which will not remove the natural oils from the surface of your skin. For skin that is very dry, use no soap at all, but cleanse your face with a cream, removing it well with a slightly damp cotton ball.

Every few weeks, use a "scrub" on your face and throat. There is no need to buy expensive products for this.

One of the best is a simple home-made scrub made of corn-meal. Just moisten a cup of finely ground, uncooked, corn-meal and rub it all over your face and throat, with your fingertips, using circular motions. This will remove dead skin cells and smoothe your skin. Remove the cornmeal with cool water.

Keep as much water on your skin as possible by the use of moisturizers. Apply when your face is slightly damp, so that the moisturizer traps the moisture beneath it. Water applied with a plant mister is a good way to pre-moisten your skin. Many moisturizers also contain sunscreen, a doubly good way to protect your face.

As for cosmetic brands, the price usually has no bearing on their effectiveness. The simplest cleansers and moisturizers are often the best.

Sun Screens. Regular use of a sun screen is recommended. Many cosmetics and makeups now contain a mild sun screen. For times when you know you will receive an extra dose of sunshine — at the beach, a picnic or on your daily walk — use a stronger SPF (sun protection factor) rated product. Top it off with a hat or visor and you will be providing your skin with the protection it needs.

In the past, a suntan symbolized "the good life" — beauty, leisure and affluence. Sunbathing was universally popular. We now know it to be a dangerous practice leading to prematurely old, leathery skin and skin cancer.

In summary, the best way to have a youthful, glowing skin at any age is to avoid sun exposure, use sun screen on a routine basis, and drink plenty of water.

Cosmetic Surgery ("Tucks & Sucks")

If you have decided that nature and your best intentions need an assist in the form of facial surgery or body contouring to improve the evidence of aging, plastic surgery may be the answer.

It is important for you to know that this is a major decision. The "nips and tucks" vocabulary used to describe some cosmetic surgery procedures give a false impression of their seriousness and the possible risks associated — not to mention the cost, which is seldom covered by insurance. (Cosmetic surgery is usually considered to be elective or non-essential surgery by insurance companies.)

First, be clear on your reasons for doing this and your expectations. Having a "new" face will not guarantee a new life. Your goal should be improvement, not perfection. Cosmetic surgery will change your appearance and remove some of the signs of aging. And it may renew your self-confidence. But, it will not automatically bring romance into your life or get you a job.

Become as well informed as possible about the procedures that are available. Read all you can and talk with women who have had the same procedure as the one you are considering.

Most importantly, select your surgeon with care. Make sure that he or she is appropriately licensed and credentialed. For plastic surgeons, the physician should be a member of the American Society of Plastic and Reconstructive Surgeons and should be certified by the American

Board of Plastic Surgery. This means that the surgeon has had advanced training and has passed an exam given by the specialty board recommended by the American Medical Association. Not all plastic surgeons who advertise have met these criteria.

If possible, have a thorough interview with at least two properly credentialed surgeons and make your decision based upon the results of your interview. As in all doctor-patient relationships, the "chemistry" between you and your doctor is important.

Some procedures will require a stay in the hospital, others may be done in an outpatient surgery facility or in the surgeon's office. Almost all will require some kind of anesthesia, a time of recuperation from the surgery and the often-associated swelling and bruising.

Here are brief descriptions of some of the most common cosmetic surgery procedures that are frequently called "tucks & sucks".

Face Lift (Rhytidectomy). This procedure removes excess or sagging skin from the face and neck.The procedure is often done under local anesthetic which numbs the area being treated. If preferred, a general anesthetic can be used, in which case you will be asleep.

Incisions are made inside the hair line and sometimes in front of the ear and under the chin. The skin is pulled up and backward and the excess skin is removed. Fat may be removed and sagging muscles tightened. Often, eyelid surgery is performed at the same time, removing excess skin and fat from the eyelids and around the eyes.

The procedure usually lasts between two and four hours, or longer, depending upon the extent of the surgery. The cost will vary depending upon the area in which you live, the extent of the surgery, the type of facility where you have the operation and the type of anesthetic you receive. A typical cost for a face lift can vary between $2,000 and $10,000 or more.

Following the surgery, bandages will be applied and sometimes a drain tube may be inserted at the back of the ear to collect blood and secretions. Postoperative pain is usually minimal and can be controlled by medications.

The surgeon will make every effort to minimize scars by making the incisions along natural crease and fold lines. Most of the scars that are created by the surgery will fade gradually and become fairly imperceptible. (Many women report excellent results from applying vitamin E to the scar line as it heals. Just break open the vitamin E capsule and massage the oil into your skin.)

If you are going to have a face lift, your objective should be a more youthful appearance, not a new face.

Eyelid Surgery (Blepharoplasty). This operation removes excess skin and fat from the upper and lower eyelids, which gives a tired or angry look and can sometimes even interfere with vision.This procedure can also do away with those "bags" under the eyes.

Blepharoplasty is often performed under local anesthesia, combined with medication to help you relax. Incisions are made in the natural creases of the upper and lower eyelids and may extend into the squint lines or "crow's feet" at the outer edge of the eye. Excess fat around the eyes is removed and excess eyelid skin is removed, as well. The hairline scars from these tiny incisions usually fade within six to eight weeks, but sometimes remain visible beyond that time.

Cold compresses and keeping your head elevated will reduce the swelling and bruising which occurs.The sutures will be removed about a week after surgery. Eye rinses and drops may be recommended by your surgeon. Because there will be some swelling and a bruised look for the first few days after surgery, you should plan to have a recuperative time of several days before returning to normal activities.

Fees for this surgery range from $1,000 to $4,000 or higher, depending upon where the surgery is done and the complexity of the operation. In most cases, this will not be covered by insurance unless your vision was impaired.

Your objective in having eyelid surgery should be a younger, more rested appearance. This surgery will not eliminate all signs of aging, nor will any other.

The "Tummy Tuck" (Abdominoplasty). Performed to remove excess skin and fat and to tighten the abdominal muscles, this operation is considered major surgery and is usually performed in a hospital under general anesthesia. Modification of this operation may be done in an outpatient setting if a less extensive procedure is performed.

The incision is most frequently made just above the pubic hair line. The loose skin and tissue is removed, making a firmer abdominal wall and tightening the muscles.

The navel is then relocated to its proper position. Some form of external compression is normally used, similar to a medical girdle.

Medication will be ordered for the soreness which follows this surgery. You will likely stay in the hospital for two or three days, then follow a restricted activity regimen for another one to two weeks.

The cost for this surgery varies widely, depending upon the severity of the operation and how long you remain in the hospital.

Your goal in having this surgery should be a flatter abdomen and strengthened abdominal muscles. It will not cause you to lose weight from other parts of your body.

Liposuction. The removal of localized fat deposits by suction can create a different body contour with relatively little scarring. The best candidates for this procedure are those who are in good health and have good skin tone. Liposuction can sometimes remove the fat that has resisted dieting or exercise.

This procedure is done by making small incisions in the skin and manipulating a suction device to loosen and remove the fat. It can be done on most areas of the body where fat collects, such as upper arm, hips, abdomen, buttocks and thighs as well as areas of the face and neck.

Costs vary widely, depending upon the extent of the procedure.

Liposuction will not cure obesity or "cellulite", but may be successful in recontouring certain parts of your body that have been resistant to other methods of fat loss.

Breast Surgery. Surgery to either increase or decrease the size of the breasts is available. For breast augmentation to increase the size of small breasts, or for reconstruction following surgery for breast cancer, a variety of methods are used involving implants of various types.

Surgery to increase the size of the breasts or to reshape sagging breasts is one of the most commonly performed aesthetic procedures. It may be done under either general or local anesthesia in the surgeon's office or in a hospital or outpatient surgical facility.

An incision is made in the area of the breast where it meets the chest or, sometimes, just below the areola (the dark area surrounding the nipple) or in the armpit. A pocket is formed beneath the breast tissue and the implant, usually a plastic envelope containing a silicone gel or saline solution, is inserted. The incisions are closed with sutures (stitches) which are removed in about a week.

Following the surgery, the breasts may feel quite firm. Many surgeons will suggest massaging the breasts to promote circulation and soften the tissue. A loss of nipple sensitivity is common but usually returns, at least partially.

Complications from this type of surgery are rare, especially with the new implants, but should be noted. Occasionally, excessive scar tissue may form or a marked asymmetry will occur — either of which may require further surgery to correct. Infections, which are not common, are usually easily treated.

Recuperative time depends upon the extent of the surgery and your own recuperative powers; two to three

weeks might be anticipated during which time discomfort
due to the surgery gradually diminishes.

Compensation by insurance is not usually available
for breast augmentation surgery. The cost may be approximately $3,000 depending upon the type of surgery, the type
of anesthesia and the facility in which it is performed.

Breast reduction (reduction mammaplasty) is performed to reduce breast size and to relieve pain in the shoulders, neck and back caused by the weight of the breasts.
Recuperation and costs are generally more than those mentioned above since the surgery is more complex.

Cosmetic Dermatology

Dermatologists, trained in skin therapy, are also often able to treat the effects of aging. Once again, be sure to
choose the physician carefully, making sure that he or she is
"board certified" in the specialty.

Wrinkling may be treated by chemical peels or injections. Other skin problems may be treated by dermabra-
sion, cryosurgery or laser treatment. Excessive hair may be
removed by electrolysis or waxing. And the superficial "spi-
der veins" in the legs, so common in women, can be in-
jected with a solution to cause them to shrink and close
(sclerotherapy).

Many medical treatments, touted by the media as
miracles for women, are not only not miracles but carry

with them certain risks. For example, Retin-A, a drug used in cream form to treat acne has also been found to minimize the fine, shallow lines caused by sun damage. However, it will not remove the deeper lines and can produce harmful side effects in some people, primarily skin irritation and an increased susceptibility to sunburn.

Before making a decision for or against a medical or surgical change in your appearance, consider your reasons carefully. Midlife women have earned their wrinkles, gray hair and less-than-svelte bodies and are beautiful as the unique individuals they are. Society's expectations (or our perception of them) seem to ask us all to look a certain way when, actually, all women and men are beautiful — small, big, young or old.

Relationships

Loving relationships don't just happen. They require time and effort. Whether with your partner, family, friends or community, treasure them. They are your "mental tonic".

If you have a partner of many years, count yourself lucky. For that relationship to have worked for so long, you have both put a lot of energy into it. Neither of you are the same people you were when you first met and fell in love. You have both grown through the years and it has required effort not to grow apart.

Research shows us that married people are healthier than those who live alone. They have fewer illnesses and signs of aging. Marriage provides a constant source of affectionate touching — a natural healer. Having a confidante and someone with whom to share life's ups and downs makes the journey easier.

But, in order to provide these benefits, the marriage must be healthy and vigorous. Each partner's dedication must be to the relationship and to the growth of the other partner, rather than to one's own selfish interests. Don't ever take your marriage for granted, no matter how long it has been in existence. Keep your marriage a honeymoon. As Victor Hugo's aging lover is said to have remarked, "There are no wrinkles in the heart".

A mutually supportive relationship with all members of your family permits you to give and receive love. Being around children and young people is important in staying young — youthfulness is contagious.

The love a parent feels for his or her child is beyond words and creates both the gladdest and the saddest hearts. I am often privileged to be present as a baby is born. The bonding that takes place between the parents and their new child is a joy to behold. Mark Twain's epitaph for his daughter (who died at a young age) is one of the most moving and poignant tributes to that parental love:

> "Warm summer sun, shine kindly here;
>
> Warm southern wind, blow softly here;
>
> Green sod above, lie light, lie light —
>
> Good-night, dear heart, good-night, good-night."

Cherish friendships. True friends, those that love us in spite of ourselves, are a gift beyond measure. As is said in Ecclesiastes: "A faithful friend is the medicine of life".

Your relationships at work also can either keep you young or age you. A supportive environment at work is as important as it is at home. Studies show that people who have positive work relationships have ten to twelve times fewer illnesses than those for whom work is a trying ordeal without support. When people who work together have a shared set of values and goals and a shared vision, they are inspired and energized by each other.

Another kind of relationship not often considered is that we are all members of a community, whether it is our neighborhood, our church or our bridge club. This community requires the cooperation of several or many people in order to accomplish its goals. As we pool our energy and skills with others, our combined efforts benefit each of us as well as the community as a whole. And you'll be building an extended base of support.

And, how is your relationship with yourself? You are the only person you will have a lifelong relationship with. Learn to love yourself — everything about you. It's the most important gift you can give to those around you.

And, don't forget about your relationship with all the others you come in contact with during this drama called life. Do you greet them with a loving attitude or with disdain or suspicion? How about the clerk in the grocery store? The person on the street or in the hallway? The driver in traffic? Do you regard them lovingly as a fellow traveler through life or do you only make an effort to be pleasant to those you know and care about? The old saying, "Smile, and the world smiles with you" is true. Reach out and greet someone with a smile and eye contact.

No relationship is entirely positive and without its ups and downs, whether it be with a spouse, your children, your co-workers or your friends. So many of us hang on to old hurts, unwilling to give them up. We can't afford to do that. Search your life and find all the resentments you are carrying with you. Forgive them and let them go. You can't afford not to. Spend your time in relationships that are optimistic and positive. Avoid those people that are chronically depressed and who complain and criticize.

The relationship that each of us has with a higher power is the one we can call upon 24 hours a day. Whether we call it God, our higher self, the inner voice, or our creative aspect — awareness of this spiritual relationship, in my judgment, is primary and essential to our well being.

As we age, we are confronted with a myriad of physical and psychological challenges. Our skin has lines that weren't there when we were twenty. Strands of gray appear in our hair. And, what in the world are those brown spots doing on our hands? Many women find that the golden years aren't really so golden after all if our mate is no longer here to enjoy them with us and our job requires more energy than we seem to have. Some would say, that's what life is all about, change as an opportunity for growth.

I'm reminded of the story of the lobster. The lobster grows by shedding its shell and growing a new one. As the hard shell comes off, the lobster is very vulnerable, covered with just the thin pink membrane that will become the next shell. It is an easy mark for predators and can be destroyed if thrown against a coral reef by an angry sea. The lobster really must risk its life to grow.

We all must take risks in order to grow. It's easy to keep doing the same old comfortable things, feeling safe in our old shell. But by taking risks, by utilizing our hard-won maturity and freedom from many of our earlier concerns, we can experience the wonder of new experiences and adventures.

The fact is that, at midlife, we have at least one third of our productive lives still ahead of us, and knowing what to expect and how to maximize our options can make these years our best years ever — a time to grow, not decline.

The change of life is often a change for the better.

Longevity is now a biological reality. Nearly every day on the Today show, Willard Scott wishes a happy birthday to at least one individual who is more than 100 years old. And these are not people who live in out-of-the-way parts of the world, but right here in our country. As Charlie Smith said, at age 134, "If I'd known I was going to live this long, I'd have taken better care of myself".

It's important to plan to be productive and to contribute to society throughout our lifetime. If you need inspiration, consider that Will Durant completed his five volume History of Civilization when he was 89; Michaelangelo designed a church when he was 88; Horowitz returned to Leningrad to give two concerts when he was 81; and Verdi composed Falstaff at 80 years of age. No one told them they must retire and stop contributing at some arbitrary, preordained time. (And I wrote this book at age 54!)

Chapter 9

WOMEN'S OWN STORIES

Telling It Like It Is

"*To be authentically alive is to be in transition*"
(Unknown)

ho are they, these women whose stories you are about to read?

They are my friends, and yours. They are our sisters, mothers, daughters and co-workers. Some of their stories speak of PMS, some of menopause. Some of their names were changed, some were not. The words are all their words — unchanged and unedited.

It is their hope, and mine, that their words will speak to you and that you will gain from them. You may find, too, that your partner will benefit from reading these stories because it will help him to understand a part of your life that he, as a male, can never experience.

These stories have been very meaningful to the women who wrote them as well as to me as I read each one and felt moved by both the candor and diversity of what was said. If you have a story, I would love to read it. You may write to me in care of the publisher:

Family Publications
P.O. Box 940398
Maitland, FL 32794-0398

Anne (Age 30)

I am the mother of two children, 8 and 4. I discovered I had PMS when my youngest child was 2. Every month I became depressed for several days before my period and always seemed to have the same feelings of inadequacy as a wife and mother. I also had powerful food cravings that seemed to make the situation worse. Then my period would start and everything would seem normal again. I did not know much about PMS at the time, but at my sister's suggestion, I read about it. Several months later I spoke to my doctor about it. He explained what he knew and suggested dietary changes and vitamin supplements. I think the biggest relief, though, was knowing that I wasn't going crazy, that there was a reason for the confusing feelings that I had during those few days a month.

What is most interesting to me is that although this happens monthly, I'm not aware of the cause while it's happening. It's not until my period starts that I realize it. I am trying to understand my body, and learn to recognize the situation. Now that I have taken the steps that I can take, I am concentrating on helping my husband to understand how PMS affects me.

Katie (Age 34)

I know very little about PMS except that it creeps into my life every month without fail.

The existence of PMS has only recently been acknowledged so the memory of its beginning in my life is vague at best.

The experiences I have had with PMS are at times self-destructive. I cannot control my emotions at this time in my cycle. The emotion most difficult for me to suppress at this time is anger. Not external or physical anger, but internal anger so intense that subconsciously I believe that in order to cope with a particular situation I must destroy a part of myself. I get depressed just getting out of bed in the morning. I dislike being in the company of other persons (does not make for good working relationships). I become insecure, very unsure of myself.

There was a particular UMS (ugly mood swing) which had a tremendous impact on my life as well as the lives of those around me. It occured at the peak of my cycle, approximately one week prior to the onset of menses. As usual my actions were triggered by a minor incident which I created into a mountain. I quit my job. My career job. How unprofessional. I was able to save face by explaining that I had put a great deal of thought into the decision to return to college and finish my master's degree on a full-time basis. They bought it and I went my merry way with no income, dependent on my husband. Bad move. The story does have a happy ending. My replacement failed horribly and I was offered my old job back with a healthy pay increase and the opportunity to remain in school and attend classes during the day.

I am under the care of a doctor now to relieve the symptoms of PMS. I have been taking Spironolactone for

the past two years. It helps, but not at peak times. Recently
my doctor suggested a prescription for a form of progesterone that is taken sublingually (dissolved under the tongue).
Taking it in this form deters nausea and is much more convenient to take than the suppository. I am on a regimen of
200 mg. morning and 200 mg. evening. The first week on it
I noticed no change. The second week on I noticed a change.
I was my old tolerant self again. Not perfect, but bearable.
There have been no side effects thus far, and the cost of the
medication I consider to be an investment in SANITY.

Joy (Age 35)

I first realized I had a serious, bona fide PMS problem when, for two days in a particular month last year, I
was unable to face the big, bad world and the people in it. I
couldn't get out of bed due to the depression I was experiencing, consequently missing work. I suffered the most incredible feelings of anxiety and tension, culminating with
the notion of wanting to kill someone. Lucky for me, but
perhaps not so for others, I had no desire to kill myself —
just someone else! While I doubt seriously I would actually
carry out such a notion, I truly believe this was at least part
of the reason I couldn't, and wouldn't, face anyone. The
anxiety — a type that is hard to explain — was just too
overwhelming.

I'd had these similar feelings numerous times over
the years, but they'd become worse as time wore on, finally
bringing me to realize the actual intensity of my problem.

In addition to the symptoms above, I suddenly was experiencing headaches (with which I normally don't have a problem), major, almost heart-wrenching crying jags, and total personality change. And, since I keep a fairly accurate, running calendar of my cycle, I was finding that these symptoms were appearing on the same days — usually ten days out from the projected start of my period. In other words, approximately on the 18th day.

It was time to do something about it. I'd made that decision after experiencing those particular two days I mentioned above. It was pure happenstance that, at the time of this breaking point, I was handed two separate written pieces on PMS (and, both from male friends!). These articles touted vitamin and nutritional supplements as steps to take for relief of PMS symptoms. Although both added disclaimers that these "programs" may not work for everyone, I thought it was worth a try and, with information culled from these articles, I began my program the next day.

The first step I took was to cut all caffeine from my diet. I started drinking coffee years ago, mainly because I worked in an office environment. One doesn't realize how much caffeine is built up in the body as a result. I stopped drinking all coffee, tea and sodas, cold turkey. All of a sudden I had, what I'm told were, migraine headaches! I was experiencing an actual withdrawal procedure that lasted approximately two weeks but, as promised, the headaches and pains went away just as quickly as they arrived. I now drink brewed decaf coffees, sometimes herb teas, and caffeine-free colas, when I want one. My diet could always use a little altering for the better, but sweets, salt and fatty foods had to be limited, and more protein, fruits and leafy veggies increased. Sounds like simple nutritional common sense, doesn't it?

I feel the elimination of caffeine in my diet coupled with the intake of calcium supplements (1000 mg., taken with magnesium for absorption) are the two winners in relieving my symptoms. Other vitamin supplements I added to my daily routine are as follows: Vitamin B-6 (300-400 mg.); Vitamin E (DL- not D-; 400 mg); Vitamin C (1000 mg); and a Centrum or multi-vitamin. Vitamin B-6 has been linked with alleviating mood swings, nervous tension, irritability and anxiety, all symptoms I experience during PMS. Vitamin E helps decrease bloating and breast tenderness and Vitamin C, among other things, helps alleviate cravings, fatigue and increased appetite. Vitamin B-6 also helps with these particular symptoms. Magnesium, as mentioined above, does more than just help with calcium absorption. Its roles include relieving feelings of depression, confusion and crying and, those suicidal tendencies. Some researchers believe that when we crave chocolate as a PMS symptom, we actually may need the magnesium in the chocolate!

Finally, exercise is all-important in relieving PMS symptoms, not to mention the other wonderful effects it can have on your body! Release of those endorphins is a fabulous stress-reducer! Whenever I feel my anxiety getting out of control, a good workout helps keep it in check. My routine includes a Nautilus workout three days a week and 30-50 minutes of aerobic exercise, three to four days a week. I either walk fast, ride a bike (mobile or stationary), or row. Sometimes I break up the 50 minutes by doing all three. You must work to get your heart rate up and sustain it for as long as possible — believe me, it works!

You may want to know how long it takes to acquire relief after beginning such a program. Well, with my vitamin supplementation alone, I felt relief by the time my next cycle began. And, while I'm still experiencing relief from my PMS symptoms, I'm not going to tell you my PMS has disappeared. It's always there, but it needs to be kept in check, always looked after. Some months I feel no symptoms at all, and some months I experience a twinge of anxiety or other symptoms that are seemingly ever-ready to jump in and take control if I let them. The trick is to do everything you can to maintain control — in other words, if you want relief from PMS symptoms, take steps to DO SOMETHING ABOUT IT! And, good luck!

Lisa (Age 38)

I am a businesswoman who thrives on stress. I feel my best when I'm busy and challenged with fifteen things happening at once.

But for the past few years, every month about a week before my period begins, I become very anxious and nervous. My co-workers tell me I'm impossible to get along with and that I "need help". I've been passed over for promotions twice that I know of because of my alleged moodiness and unpredictability. I know my mood swings and angry outbursts are due to PMS, but knowing it doesn't help me to control it.

Sometimes I have difficulty breathing because my throat feels so tight and I find I'm either holding my breath or taking little shallow breaths. The feeling of not being able to get enough air is terrifying. I've read about panic attacks and think that must be what happens to me.

I'm trying to improve my diet and plan to take a stress management class. And now that I know about the PMS support group, I'll go to that. Maybe the other women can tell me how they manage to cope!

Esther (Age 46)

I suppose I've suffered from PMS ever since I began menstruating; I just never knew it. All the same, I always knew that something wasn't right. For five to seven days before the start of my period, I was like a 45 rpm record playing at 78 rpm speed. My symptoms were: sleeplessness, depression, lumpy breasts, craving salty and/or sweet foods, edema, and an irritability that would escalate into progressive bitchiness. The symptoms grew worse with each passing year.

When I had a partial hysterectomy due to excessive bleeding, I was sure my troubles were over. Little did I know they were just beginning. While my symptoms had been uncomfortable and unpleasant, they were for the most part manageable. But, several months after my hysterectomy, my symptoms began escalating so quickly that I felt like I was on a roller coaster ride gone wild. For three weeks of the month I was a rational, loving human being, then for one week, something as simple as a glass left on the table could turn me into a frenzy. I think the scariest part of it was, I could almost step outside of myself, and watch this crazy woman yelling and screaming. And I didn't know how to stop it. I felt totally out of control. I would binge, eating sweets, then chips, then back to sweets again with reckless abandon. When depression would hit, there were times I felt suicidal. My gynecologist said it was all in my head. I thank God I had the sense to switch doctors.

I'm now taking estrogen by patch and progesterone 400 mg by suppository during my cycle. I use one or two a day for 12 days, depending on how I'm feeling, then I stop for 14 days. Not all of my symptoms are gone. I still get lumpy breasts, have problems sleeping and crave sugary and salty foods, but the end result is that now I'm back in control.

Eleanor (Age 67)

My doctor started me on estrogen (Premarin) when I was 44. I had been experiencing hot flashes and vaginal dryness. This was a definite help for me and relieved my uncomfortable symptoms.

After 12 years, my doctor changed my therapy to shots of a combination of estrogen and testosterone administered once every three weeks.

I must say that hormone therapy has made a big difference in my life's quality— especially my sex life.

Louise (Age 50)

Who would believe that I could be sitting here, dressed in my beautiful Mexican kaftan, gourmet dinner roasting in the oven, gorgeous table set for our guests, sipping a glass of wine, and I am dripping wet from head to toe. I'm soaked. My body temperature must be 120 degrees. I'm trying to be gracious, elegant, the perfect hostess while rivers of sweat are pouring down the inside of my thighs. My breasts are drowning, my cheeks broiling, every crease in me is aflow, my body and the room temperature are rising by the second. Now I can't catch my breath (am I suffocating?), now it's faraway sounds of voices while I struggle to focus my attention so they won't notice that I'm gone, washed away in a pool of sweat and heat.

This, I tell myself, must be a "Hot Flash", but why me, God, a mere 50 years old with the mind and spirit of a woman half that age. Maybe if the orthopedic surgeon I saw last week hadn't assumed I had already experienced menopause (my grey hair led him to that conclusion) this never would have happened! Another funny — this new happening in my life reminds me of when I first got my period. I was equally surprised, embarrassed and secretive about it then, despite the books, belts and napkins that I had practiced using for two years before. The inevitable monthly red-stained panties, skirts and bedsheets; the cramps, the uncomfortable bulge in a veteran tomboy's jeans were all too much for me. Now, 38 years later and I'm reduced to the same feelings of embarrassment, discomfort and body going nuts on me again.

Now I have entered the ranks of my mother's bridge club — menopausal female is another description to add to my list of identities. A sign-post into a new phase of my life — elder, crone, wise woman, old lady, biddy, grandmother— all these await me — ushered in by a 21-gun hot flash!

Marion (Age 64)

I have been taking hormones for 25 years, ever since my hysterectomy.

But now I've been taken off them since my surgery for breast cancer. I never had hot flashes while I was taking the hormones. Now I feel like I'm on fire all the time. I sleep with a fan on and my bed is soaked every morning. My doctor said I've got to put up with it since I can't take any estrogen. "No hormones for you", he said.

I'm very discouraged about it, but guess I'll just have to learn to live with it.

Ginger (Age 59)

I don't remember any real trauma upon turning 50 years old. It was a lovely birthday party and in the photos I look happy and rather young to have hit the half-century mark. In fact, I felt healthy and young, too. I have been blessed with good health.

It was the spring before turning 54 that, at my semi-annual check-up, my gynecologist said it's time you start taking estrogen. What on earth, I thought, I'm still as regular as clock work. But he had foretold the happenings of that late summer when I suddenly stopped my periods. There was no physical trauma involved. I just stopped. I took the estrogen for about a year, then, after reading some of the "scare" literature which is packaged with the pills, I stopped taking it. About six months later, I started taking estrogen again with the progesterone and have been taking it for three years.

I'm not sure that I really want to do this. Also, in my early 50's, I started taking thyroid pills for a slightly under-active thyroid gland...I don't like to take these either. Now the person I always thought so healthy is a "chemical woman"! I really have trouble communicating this to my doctor. I've been his patient for about twenty years and should be able to tell him how I feel. Not long ago, I stammered out that I don't enjoy sex the way I used to...so he gave me a shot to make me "sexy". I guess it did engorge my clitoris for about a month...but it also gave me hair on my upper lip! Not worth it, I'd say! I don't particularly want to change doctors but I wish I could get clearer answers to my questions. Like, for instance, how long will I be taking hormones? Is it really necessary if I haven't been diagnosed as a candidate for osteoporosis? I just pray that research doesn't reveal that it really is dangerous ... actually I worry more about the blood clots, cancer and my stomach than I do about my bones. I feel some clear answers are needed about hormones ... and I know my communication with my doctor should be better.

Sindi (Age 46)

Due to a great deal of stress in my life, I experienced premature menopause at the age of 36. At first it was great because of the absence of menstruation. (The phenomenon that every little girl looks forward to starting as she approaches womanhood, and every woman looks forward to ending!)

At about age 40, the classic symptoms began; night sweats that woke me up drenched in perspiration, anxiety attacks, and insomnia, to name a few. I sought the help of my gynecologist and estrogen and progesterone were immediately prescribed, without testing my hormone level to determine an appropriate amount.

After almost five years of being menstruation free, I did not relish the idea of starting that up again. But, as it turned out, that was the least of my worries.

Some women gain weight while on hormones. I became very depressed. But, worse than that, I began not only to have a monthly period to look forward to but, severe cramping. It was like I was in labor and the baby was being ripped from my body. Along with the labor-like pain, was nausea, perfuse sweating and extreme weakness. I could count on this every month until I had a good menstrual flow. The last time I took the progesterone, this happened twice. At that point I determined that it just wasn't worth the pain I had to endure and I stopped taking the hormone medication.

I must say, too, that the only response I got from my doctor was very negative, saying that there couldn't be a connection. I found that hard to believe. Since I stopped taking the hormones, the pain has never returned.

Alice (Age 55)

My meopause began abruptly at age 45 when I had a hysterectomy with removal of my uterus and ovaries. A D&C had identified a "pre-cancerous" condition which was causing a steadily worsening bleeding problem.

I recall going to my doctor's office for the pre-op visit and the nurse said, "Oh, you're going to just love your hysterectomy!" I thought she was insensitive and unfeeling to say such a thing, since I was very upset over the prospect of the surgery.

Ten years later, I have to say that she was absolutely right. I've never had the first problem as a result of the surgery. And, since I began hormone replacement immediately, I haven't had the usual menopausal symptoms. The best part was the transition from heavy, flooding periods to none at all — a joy to one who loves to travel. No longer did I plan trips around my menstrual cycle or travel with an extra suitcase full of tampons and sanitary pads. And, maybe best of all, since there is no concern about the risk of endometrial cancer from taking estrogen, I am not required to add progesterone.

I have felt great during these last ten years, with energy in abundance and no physical problems. This part of my life is, in many ways, the best part ever.

Jeanine Suelyn (Age 48)

These years of change have been a strange mixture of confusion, intensity and peace. Premenopause, menopause and postmenopause have all occurred for me between ages 38-48.

When the process began for me, I experienced skipped periods, continual bleeding for two to three weeks or spotting for a month at a time. This time was an emotional roller-coaster. I was physically uncomfortable part of the time and the "normal" intensified emotions I had experienced throughout my life, before and during my normal

menstrual periods, were ever-present during this premenopause phase.

Menopause, in its "full blown form" brought hot
flashes, night sweats, free-floating anxiety, mood swings and
sleepless nights. I thought at times my body thermostat had
turned off at 102 degrees and the sleepless nights were provided for my intellectual stimulation. The hours between
1:00-3:00 a.m. existed for the sole purpose of my reading
everything that had ever been published on hormone replacement therapy, menopause, natural therapies, herbs,
vitamins and the power of positive thinking. I had decided
early in my life that I was going to enjoy and experience
every part of my female life from a natural, normal way.
My grandmother was the most wonderful person I had ever
known and she had 10 healthy kids, all by natural childbirth. When I asked her about menopause, she said there
was nothing tough about it; she just went through it naturally. She was sick for only two years of her life, the last two
years. She died when she was 86.

I also decided to not have MENopause, but instead
to have WOMANpause. This meant that I was simply PAUS-
ING to fully experience all of my womanness, all of my
emotions, dreams and body sensations in a few compacted
years. Depression showed up more often than in the past
and at times it took very little for me to cry or become very
"pissed off". Talking to my women friends going through
the same process and some that were years beyond it really
helped me to keep a sense of balance. When I became really
upset, they were a "reality source". We helped each other
identify "hormone thinking". During "hormone thinking",

we gave each other more support and laughter, promised to make NO decisions until the feelings passed away and told each other "take two jelly beans and call me again in the morning"! My doctor told me again and again that he could make ALL OF THIS go away with a few magic pills called hormones, but I would not agree since my mother had cervical cancer within one year of going on estrogen. This, at times, felt like a battle over my body.

I am now in a phase of no night sweats, hot flashes, sleepless nights or breakthrough bleeding and my emotions are even. And I feel joyful most of the time. A sense of peace has settled in and I can really say that I am glad I have done this without any anti-depression drugs, fluid loss pills, sleeping pills, wake-up pills or hormones. Now that I have not had a period for over a year, my nights of reading have led me to believe that using hormones at this phase, post-womanpause, will be the safest time for me to begin.

Carla (Age 67)

There's a title for this letter: "How to Have Sex Without Even Crying". It was also my motivation for the events that follow.

Sex, in the beginning of my menopause, was great and free. No more birth control medications or preparation. I was on estrogen therapy for about ten years after, but stopped for a couple of years. It was then I began noticing the vaginal dryness. I tried to alleviate this with estrogen cream, but had a severe burning reaction. I then returned to estrogen by mouth with still vaginal discomfort. I changed doctors and the new one prescribed twice the dosage and things seemed better. After a year, I had a breakthrough of bleeding and spotting. The doctor examined me and said to drop to half the estrogen I was taking. I was still uneasy and went to another gynecologist. He was quite alarmed by the extreme thickening of the walls of my uterus and immediately scheduled me for a D & C, which happily showed no sign of cancer. He then explained that I needed the estrogen for the dryness and also to stave off any signs of osteoporosis. However, it was necessary to take progesterone along with the estrogen to keep the buildup on the uterus from occurring again. I would also have a period each month.

As soon as the D & C was checked and I had healed I began that therapy. Never have I had such misery. The first month the period was excessively heavy for at least three days. I could not leave the house as I was saturating two and three pads at a time. I called the doctor and was told to continue as it would slow down as my body got used to the medication. The next two months were worse with cramps and heavy bleeding. One night I woke up to find our bed saturated from this heavy flow. We had to get up, change the bedding before we could get back in bed and then had trouble sleeping for fear it would happen again. I felt I was a prisoner of my body, unable to leave the house for those days.

There has to be a better way. On informing the doctor that if he had no solution to my problem with this medication, I would stop it all. This I have done. My active life-style has returned unhampered, but sex is impossible and painful.

Bobbe (Age 50-something)

Humor and hormones ... they seem to go hand in hand. I remember when I first encountered hormonal action in my adolescent body. A message was released from my pituitary gland and sounded in my brain — RED ALERT! — BOYS! What fun! It felt good.

Some time later, I fell seriously in love and got married. More fun. It still felt good.

I guess the funny part was the hormonal effect felt while I was pregnant, (meaning funny as in strange, not funny - haha). Still, it was fun. I could eat what I pleased (eventually) and it still felt good. So good that I did it five times!

The nicest thing my hormones ever did for me is kick in while I was lactating. Thank God for prolactin, the "feel good" hormone. It made nursing those little Lyon cubs a joy, almost euphoric. Guess that's why I nursed for so long. I haven't felt that good before or since.

And now the BIG test — menopause! I must confess I've managed to be spared the "hot flashes" (my hormones' way of saying goodbye?). All I've really noticed is some crazy mood swings and occasional lapses of memory. One of my daughters has dubbed this time of my life my "mental pause". I like that, it feels good and I'm still having fun.

Now two of my daughters are pregnant and the third one is getting married. My two sons have yet to encounter the joys of parenthood. I could swear those hormones are kicking in again. I still feel good and I'm still having such a lot of fun. Thank God for humor and hormones — how on earth would we ever survive without either?

P.S. Those memory lapses are a sign of genius.

Hilda (Age 82)

I was 40 years old when my periods stopped. "Happy Day", I thought. In my ignorance I did not know what was in store for me. I went to see a doctor and he put me on estrogen. I was loyal for about six months and then stopped taking them altogether. I was leading a very busy life. Little did I know what the outcome would be.

I began to have backaches and went to a chiropractor. That didn't help at all. I tried many other sources but to no avail. The backaches grew worse.

I tried other doctors and tried a half a dozen different things, but nothing helped. The back pains increased in intensity. I tried a new doctor who gave me three shots a week. After a period of six months and many different treatments, nothing helped. Today my back is worse. I am in constant pain. It is becoming difficult for me to be as agile as I once was.

I learned my lesson well and am paying dearly for my neglect to take medicine in the early stages of my osteoporosis.

Anne-Marie (Age 43)

My puberty was a time of self-initiated, although unconscious, suppression. Shortly after I had my first period I developed a relatively mild case of anorexia, by which I suppressed my periods for about 4 years, through dieting and weight loss. At about age 17, I required hormonal stimulation of my periods, after which I menstruated regularly. At that time I was no longer engaging in anorexic behavior. Why did it seem vital to me to control my appetite for food? Was this a symbolic way of controlling my sexual appetite?

Looking back, at age 43 to the girl I was at 13, I can see that the reasons I "shut down" my development through high school were: lack of sexual education and knowledge, lack of contraception education, and lack of "permission" from parents, teachers, or peers to experience myself as a sexual being. Two years prior to the development of the anorexia I was struggling intensely to control a developing sexual interest which I had been led to believe would result in eternal condemnation. At age 11, I began to be aware of sexual thoughts. I'm sure these had been present as a younger child but that I had not been aware of them. My

Catholic teachers led me to be afraid, indeed terrified of these thoughts, reflecting their own basic fears of sex. So the stage had been set for me to experience my body as "bad" early in my life.

Sexuality was never discussed in my home, although my parents were affectionate and had an ongoing sexual and romantic relationship while I was growing up. My father, embarrassed, explained the "facts of life" to me on one occasion and little more was said. But long before I knew the mechanics of intercourse, my mind had been set against sexuality by my religious training and the silence in my home about sexual matters. Attempting to control or stop sexual thoughts is a futile pursuit. Thus it was quite a relief when I found out I could successfully control my weight. My sexual thoughts became less of a preoccupation while my concern for my body image became more consuming, in a typical anorectic fashion.

I dated and had boyfiriends, going through "normal" adolescent experiences. These dates never involved sexual feelings or more than a goodnight kiss. I began to menstruate at 17, and delayed having sex until I was 21 years old. It was soon after my 21st birthday that I became sexual with a man for the first time. As an independent adult, I felt suddenly free to make choices which were not in accord with my upbringing. My anorexia was short-lived but the suppression of myself has been a strongly ingrained tendency against which I have struggled with increasing success.

From the time I was 13 until today I have always been strongly controlled concerning my eating. To others I seem unusually well disciplined in terms of my food preferences and avoidances. I have a relatively healthy diet, and have maintained myself in good physical health and conditioning. My reproductive system has functioned healthfully, even though I would again suppress estrogen production with birth control pills over a period of 10 years. I feel well integrated sexually and enjoy sex in the context of a loving relationship.

My experience has led me to feel that we should be more open about sexuality in our schools and homes. Today our media is very explicit about sexual behavior of a type which most often degrades women, but for the most part our teachers remain silent, with sex education still a controversial issue. Given guidance and education in the prevention of pregnancy and disease, and open communication with parents, teenage years could become an important time of sexual experimentation which could enable young people to make more intelligent choices about love, marriage, and family.

Menstruation marks the beginning of a woman's sexual and reproductive life and is the first dramatic encounter she has with her "hormones". As a psychologist, my interest in women's issues includes the way in which social, cultural, and biological factors interract with individual psychological factors. It is the work of society and culture to provide guidance to enable young women to experience the changes of puberty as a wonderul gift, as opposed to the "curse". Appreciation of the deep significance of sexuality in our lives as individuals and as part of the human race is the hoped for psychological outcome.

The impact of society, family and culture cannot be ignored in considering the way in which hormones function. Nor can the conscious or unconscious motives of individual women.

Glenda (Age 49)

I had turned 48 on October 20, 1988. I sat up in the bed, pulled out my bedside drawer, picked up the 38 pistol and proceeded to load it with bullets. It wasn't that I had planned to "do myself in", but ever since October of 1987, I noticed that my usual "down" days of my monthly cycle had lengthened. I seemed to be continually depressed, and for no apparent reason. I returned my attention to the pistol I was holding, and asked myself, "Do I really want to do this to myself and those that were trying to help me? I might as well shoot my husband, my two sons, and my mother, because I would be doing just that, .. if I shot myself!" And then an additional thought occurred to me, "I will probably botch this job, blow my ear off, AND be unable to hear the tennis ball bounce during my weekly games!"

Being only one of a million or more menopausal women in 1989, I would like to tell you my story, in the hope that women AND men may find solace, helpful similarity, and understanding of the unique feelings and struggles that females experience daily.

As my periods of depression continued to lengthen, I talked to my internist/cardiologist. Over a period of a year or more, he prescribed one shot of estrogen, and in subsequent visits he prescribed Premarin; then added Medroxyprogestron; Amitriptyline for depression; and eventually made an appointment with a friend of his that was a psychiatrist. The psychiatrist determined that I was suffering from major depression, and suggested "shock treatments" as one alternative, since I was already on hormone therapy and medication for depression. I decided against the shock treatments and hoped the medications would eventually improve my mental attitude and depressed condition.

Throughout the summer, I continued the hormone medication, and the antidepressant, but neither seemed to effectively improve my condition. During this time, I also saw several counselors, a female gynecologist, my regular chiropractor, friends having similar feelings and my husband's internist. Sometime during this year, it was suggested I have a complete blood test. Both internists discovered I was hyperthyroid, and the thyroid medication I had been taking for twenty years was then reduced by one-half. My husband's internist also took me off all hormone medication, and reduced the others to the minimum. My depressive, moody days seemed to improve slightly, but by this time I was totally convinced that I was the "guinea pig" for all menopausal women. Therefore, I contacted a friend of mine that had recently had similar emotional problems, and she directed me to a specific psychiatric group that had helped her. This psychiatrist took me off all medication; put me in the hospital for a week, and diagnosed my condition as bipolar disorder. I am presently taking .01 mg of Thyroid and 300 mg of Lithium Carbonate three times a day, and my mood swings and intense depression are at a minimum. Hopefully, this WILL continue! In retrospect, after a period of almost three years, I sincerely wonder if what happened to me was physical, mental, emotional, "burn-out", work/ family stress, or ALL of the above!! I do know that my illness almost erased my life. If it had not been for family, friends, and PATIENCE, I would not be here to tell my story. My major concern is for the thousands of women that have no family or friends in whom they can confide, or are limited financially as well as educationally in available services. In this rapid pace in which we live, the patience, love and proper medication that was provided me, was not only helpful, but LIFE-SAVING.

------------------- ❧ -------------------

As you see, women have all sorts of experiences with hormonal issues and choose different methods of coping. We each need to find our own solution; it is through education about options that we can make our best choices. The reading list which follows suggests books that deal in depth with many of the topics that have been discussed in this book. They contain important information on women and on things we can do to empower ourselves. Learning how to relate to the medical establishment is an important first step in helping ourselves. Many of the stories in this section pointed out the frustrations women felt in trying to get help. Knowing the right doctor to go to isn't always easy, but knowing your rights as a consumer of medical services is essential. I hope this book has helped, and I wish you well.

Epilogue

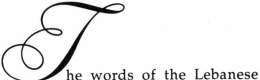he words of the Lebanese poet, Kahlil Gibran, have spoken to me for many years. These words on beauty from **The Prophet*** express my thoughts on the real beauty of us all which transcends time and the physical world.

And a poet said, Speak to us of Beauty.
And he answered:
Where shall you seek beauty, and how shall you find her unless she herself be your way and your guide?
And how shall you speak of her except she be the weaver of your speech?

The aggrieved and the injured say, "Beauty is kind and gentle.
Like a young mother half-shy of her own glory she walks among us."
And the passionate say, "Nay, beauty is a thing of might and dread.
Like the tempest she shakes the earth beneath us and the sky above us."

The tired and the weary say,
"Beauty is of soft whisperings. She speaks in our spirit. Her voice yields to our silences like a faint light that quivers in fear of the shadow."

But the restless say,
    "We have heard her shouting among the mountains, And with her cries came the sound of hoofs, and the beating of wings and the roaring of lions."

    At night the watchmen of the city say, "Beauty shall rise with the dawn from the east."
    And at noontide the toilers and the wayfarers say,
    "We have seen her dancing with the autumn leaves, and we saw a drift of snow in her hair."

    All these things have you said of beauty,
    Yet in truth you spoke not of her but of needs unsatisfied,
    And beauty is not a need but an ecstasy.
    It is not a mouth thirsting nor an empty hand stretched forth,
    But rather a heart enflamed and a soul enchanted.

    It is not the image you would see nor the song you would hear,
    But rather an image you see though you close your eyes and a song you hear though you shut your ears.
    It is not the sap within the furrowed bark, nor a wing attached to a claw,
    But rather a garden for ever in bloom and a flock of angels for ever in flight.

    People of Orphalese, beauty is life when life unveils her holy face.
    But you are life and you are the veil.
    Beauty is eternity gazing at itself in a mirror.
    But you are eternity and you are the mirror.

                                        -Kahlil Gibran

*Used with permission

Glossary

GLOSSARY

adrenal glands: small glands that sit on top of the kidneys which secrete the hormone adrenaline.

androgen: male sex hormone that develops and maintains masculine characteristics.

arthralgia: pain in a joint; experienced by some women during menopause.

atrophic vaginitis: thinning of skin in the vagina.

bone loss: the gradual loss of calcium and protein from bone, making it brittle and susceptible to fracture.

breakthrough bleeding: any visible blood when not expected.

calcium: a mineral in bone that gives it hardness.

cervical smear: also called a Pap test; a simple procedure to test for cancer during which a tiny sample of cells is taken from the cervix and/or vagina.

cervix: neck of the uterus located at the top of the vagina.

climacteric: the years leading up to and following the last menstrual period. Also called "peri-menopause".

diuretics: drugs given to increase the flow of urine.

"dowager's hump": type of curvature of the spine (kyphosis); can occur as a result of osteoporosis.

dysmenorrhea: difficult or painful menstruation.

dyspareunia: difficult or painful sexual intercourse.

endometrium: lining of the uterus (womb).

FSH: follicle-stimulating hormone; produced by the pituitary gland which signals the ovaries to develop the egg-containing follicles.

fibroids: non-cancerous growths in the uterus that occur in 20-25% of women before menopause; cause no symptoms in most women.

hormone: chemical substance which is secreted by a gland and travels by the bloodstream to other 'target' organs which it stimulates to act.

hysterectomy: surgical removal of the uterus (womb).

hyperplasia: an abnormal increase in the number of cells in body tissue.

hypertension: high blood pressure.

kyphosis: unnatural forward bending of the spine; the type caused by osteoporosis is often called "dowager's hump".

oophorectomy: surgical removal of the ovaries.

osteoporosis: thinning of density of bones due to aging and other factors; common in elderly people, especially women, in whom it often begins at the time of menopause.

ovum: egg.

ovary: gland in women which contains the female egg cells and which produces the hormones estrogen and progesterone (There are normally two ovaries, one on either side of the uterus.).

ovulation: release of an egg from the ovary.

Pap test: another name for a cervical smear; named for its originator, George Papanicolaou, who first described the use of the vaginal smear in 1928.

pelvic floor: muscles at the bottom of the pelvis which support the pelvic organs; can be strengthened by the Kegel exercise.

pituitary gland: small gland at the base of the brain which produces several hormones, including those which stimulate the ovaries to produce estrogen and progesterone.

progesterone: hormone produced in the ovary which is important for the maintenance of pregnancy.

progestogen: artificially produced progesterone.

prolapse: when an organ (such as the uterus or bladder) moves from its usual position in the body.

remodeling: a renewal process by which some bone cells add and others remove small amounts of bone.

surgical menopause: the ending of menstrual periods when the uterus and sometimes the ovaries are removed surgically.

syndrome: collection of symptoms.

testosterone: male sex hormone produced in the testes.

urethra: the tube that carries urine from the bladder to the outside of the body.

uterus: womb; muscular, pear-shaped organ in the pelvis, the lining of which is shed monthly, as menstruation; the organ in which the fetus develops during pregnancy.

vagina: tube-like passageway leading from the uterus to outside the body; sometimes called the birth canal.

vulva: area between the legs and around the opening of the vagina.

womb: uterus.

Reading List

READING LIST

American Medical Association. *Woman Care.* 1984.

Barbach, Lonnie. *For Yourself: The Fulfillment of Female Sexuality.* 1975.

Beard and Curtis. *Menopause and the Years Ahead.* 1988.

Benson, Herbert, M.D. *The Relaxation Response.* 1975

Borysenko, Joan. *Minding the Body, Mending the Mind.* 1987.

Boston Women's Health Book Collective. *The New Our Bodies, Ourselves* . 1984

Browne, Joy. *Nobody's Perfect: How to Stop Blaming and Start Loving.* 1988.

Budoff, Penny W., M.D. *No More Hot Flashes and Other Good News.* 1984.

Colgrove, Bloomfield & McWilliams. *How to Survive the Loss of a Love.* 1976.

Comfort, Alex. *The Joy of Sex.* 1972.

Cooper, Kenneth, M.D. *Preventing Osteoporosis.* 1989.

Cutter, Garcia, Edwards. *Menopause.* 1983.

Davidson, Joy. *The Agony of It All.* 1988.

Doress and Siegal.*Ourselves, Growing Older.* 1987.

Edelstein, Barbara. *The Woman Doctor's Medical Guide for Women.* 1982.

Fonda, Jane. *Women Coming of Age.* 1984.

Friday, Nancy. *Jealousy.* 1985.

Gawain, Shakti. *Creative Visualization.* 1978.

Greist, Jefferson & Marks. *Anxiety and Its Treatment.* 1986.

Harrison, Michelle, M.D. *Self-Help for Premenstrual Syndrome.* Revised edition, 1985.

Helmstetter, Shad, Ph.D. *What to Say When You Talk to Yourself.* 1986.

Henig, Robin Maranta. *How a Woman Ages.* 1985.

Jampolsky, Gerald G., M.D. *Teach Only Love.* 1983.

Kinder & Cowan. *Husbands and Wives.* 1989.

Kitzinger, Sheila. *A Woman's Experience of Sex.* 1983.

Kushner, Harold. *When All You've Ever Wanted Isn't Enough.* 1986.

Lauerson & Whitney. *A Woman's Body.* 1987.

Lerner, Harriet Goldhor. *The Dance of Intimacy.* 1989.

Linkletter, Art. *Old Age Is Not for Sissies.* 1988.

Masters and Johnson. *On Sex and Human Loving,* 1982.

Meyer, Roberta. *Listen to the Heart.* 1989.

Morgan, Elizabeth, M.D. *The Complete Book of Cosmetic Surgery.* 1988.

Notelovitz, Morris. *Stand Tall.* 1982.

Novick, Nelson Lee, M.D. *Super Skin.* 1988.

Orbach, Susie. *Fat Is a Feminist Issue.* 1982.

Peck, M. Scott. *The Road Less Traveled.* 1978.

Peck & Avioli. *Osteoporosis.* 1988.

Pelletier, Kenneth. *Mind as Healer, Mind as Slayer,* 1977.

Sanford & Donovan. *Women and Self-Esteem.* 1984.

Segall, Paul. *Living Longer, Growing Younger,* 1989.

Shain, Merle. *When Lovers Are Friends.* 1978.

Sheehy, Gail. *Passages.* 1976.

Simon, Sidney. *Getting Unstuck.* 1988.

Smedes, Lewis. *Forgive and Forget.* 1984.

Smith, Manuel J. *When I Say No, I Feel Guilty.* 1973.

Stewart, Guest, Stewart, Hatcher. *My Body, My Health.* 1979.

Stoppard, Miriam. *Everywoman's Medical Handbook.* 1988.

About the Author

Alice T. MacMahon R.N., M.P.H. is Director of the Center for Women's Medicine at Florida Hospital in Orlando. Although she enthusiastically accepted the challenge to develop a bridge between the traditional practice of healthcare and the women's consumer movement, Ms. MacMahon came to the position already prepared to deal with being on the leading edge of the healthcare delivery system. She was one of the earliest proponents of awake-and-aware childbirth in the South and was able to innovate and produce many such programs there. One result has been the recognition of her institution as a model for hospital-based Lamaze childbirth programs. It became and remains one of the largest institutions offering pre- and postnatal education and care with over 3000 couples enrolling in one or more of the hospital's parent education programs each year. These and other proactive programs linking women and families to the forefront of institutional healthcare have given Ms. MacMahon the opportunity to put into everyday practice her belief that the whole person is one who can manage a balance among the physical, mental and spiritual aspects of life. Her earlier book, *All About Childbirth*, is used nationwide as a guide to prepared childbirth. She is a frequent speaker on women's health issues and is a consultant on the topic of the design and implementation of women's healthcare programs.

Please see the order form on the other side of this page for additional copies of this book. Audio cassettes by Alice MacMahon are also now available.

SPECIAL DISCOUNTS AVAILABLE
ON QUANTITY ORDERS OF THIS BOOK

Special discounts of up to 25% may be available to your institution, club or other organization if multiple copies of *Women & Hormones* are ordered direct from the publisher. For orders of less than 10 copies, please contact your bookseller or send $9.95 for each copy plus shipping and handling charges as noted below. Ideal for classes and groups.

Professional Discounts are available on request.

ALICE MACMAHON NOW
"LIVE" ON AUDIO CASSETTE

Hear the warmth, wit and personality of the author of *All About Childbirth* and *Women & Hormones* as she talks directly to and about everyday women and their healthcare concerns. More than an audio version of *Women & Hormones*, Alice's talks to groups and interviews with real people add a whole new dimension to her printed messages about the needs, options and possible steps to empowerment that are now available to women. For yourself or to share.

— — — — — ✀ — — Please Use This Order Form — — ✀ — — — — — —

Date _____

BOOK:

1 - 3 copies; $9.95 ea. Send ____ books @ 9.95 + 2.00 S&H = $_____.____

4 - 9 copies; $9.95 ea. Send ____ books @ 9.95 + 3.50 S&H = $_____.____

10- 19 copies; $8.95 ea. Send ____ books @ 8.95 + 5.00 S&H = $_____.____

20- 49 copies; $7.95 ea. Send ____ books @ 7.95 +10.00 S&H = $_____.____

TAPE:

Cassette tapes @ $12.95 each. Send ____ tapes + 3.50 S&H = $_____.____

Total Books incl. S&H = $_____.____

Total Tapes incl. S&H = $_____.____

In FL, add 6% sales tax = $_____.____

(Sorry, we do not accept credit cards)　　　　　**Total Order:** $_____.____

Please send check or money order to: Family Publications
P.O. Box 940398
Maitland, FL 32794-0398
Phone (407) 539-1411

(Please Print)

Name_____

Title_____

Organization_____

Street_____

City_____ State _____ Zip_____

5M190